Hands-On RESTful Web Services with TypeScript 3

Design and develop scalable RESTful APIs for your applications

Biharck Muniz Araújo

BIRMINGHAM - MUMBAI

Hands-On RESTful Web Services with TypeScript 3

Commissioning Editor: Richa Tripathi
Acquisition Editor: Shriram Shekhar
Content Development Editor: Rohit Kumar Singh
Technical Editor: Royce John
Copy Editor: Safis Editing
Project Coordinator: Vaidehi Sawant
Proofreader: Safis Editing
Indexer: Rekha Nair
Graphics: Alishon Mendonsa
Production Coordinator: Nilesh Mohite

First published: March 2019

Production reference: 1220319

Published by Packt Publishing Ltd.
Livery Place
35 Livery Street
Birmingham
B3 2PB, UK.

ISBN 978-1-78995-627-6

www.packtpub.com

To my wife, Aline Lopes Coelho, and my daughter, Liz Lopes Muniz, for exemplifying the power of determination and love. When I started writing this book, Liz was only three months old and my wife and I were both working full-time. To this day, I have no explanation of how we have had the energy to handle everything together as the fantastic family that we are. Thank you, Aline, for being my loving partner throughout our life journey together, and a special thanks to my little one, Liz, for spending time watching me with her pacifier while I was writing this book during the wee small hours of the night.

– Biharck Muniz Araújo

`mapt.io`

Mapt is an online digital library that gives you full access to over 5,000 books and videos, as well as industry leading tools to help you plan your personal development and advance your career. For more information, please visit our website.

Why subscribe?

- Spend less time learning and more time coding with practical eBooks and Videos from over 4,000 industry professionals

- Improve your learning with Skill Plans built especially for you

- Get a free eBook or video every month

- Mapt is fully searchable

- Copy and paste, print, and bookmark content

Packt.com

Did you know that Packt offers eBook versions of every book published, with PDF and ePub files available? You can upgrade to the eBook version at `www.packt.com` and as a print book customer, you are entitled to a discount on the eBook copy. Get in touch with us at `customercare@packtpub.com` for more details.

At `www.packt.com`, you can also read a collection of free technical articles, sign up for a range of free newsletters, and receive exclusive discounts and offers on Packt books and eBooks.

Foreword

REST has been around for many years, so why should someone write a book about it now? In our book, *Building Evolutionary Architectures*, we described the software development ecosystem as a dynamic equilibrium—equilibrium because all the pieces fit together, but dynamic because any part (technology, technique, tool, framework, library, and so on) can change at any time. Given that architects and developers work in such a volatile environment, it is useful to revisit technologies that are still used on a continuous basis because we constantly figure out new ways to apply useful technologies.

REST certainly qualifies as a popular technology that continues to gain traction. Just like all other active software tools, the tools and approaches evolve over time, which makes this book a timely addition to the corpus of books relating to REST. The author does a terrific job of establishing the technology underpinning REST, but there are also chapters discussing how the use of this protocol has evolved over time, including numerous best practices regarding how real teams use REST and related tools to solve real problems.

This book is well organized, dividing its time between critical academic definitions (REST came from academia, after all) and practical advice on how to apply it. For example, Chapter 2, *Principles of Designing RESTful APIs*, provides an excellent overview of how teams use REST for different purposes, and not just how the bits wire together.

I think this book will become a valuable addition to developers' libraries. It serves as both an excellent introduction and overview, but also as a fantastic reference book for establishing best practices and other enduring advice. Congratulations to Biharck for writing a clear and compelling book on a complex subject and its related ecosystem.

Neal Ford

Director / Software Architect / Meme Wrangler at ThoughtWorks

Contributors

About the author

Biharck Muniz Araújo is passionate about technology and academic research. He has been working as a software architect and lead programmer for the past 12 years. With over 16 years' experience, he has been working with technology in relation to large-scale problems associated with web projects that demanded high-security standards for information transmission in companies in a variety of sectors, including telecoms, health, and finance. His background is in computer science, and he has a Master's degree in Electrical Engineering and a Ph.D. in Bioinformatics. Currently, he is focused on performance and algorithm design.

About the reviewer

Ruchitha Sahabandu is a software engineer, technical trainer, and mentor with more than 14 years' experience in designing and developing business applications, embedded systems based on JavaScript frameworks, JEE, Oracle, and C/C++ technologies. He has worked in many challenging projects run on Agile and TDD, SDLC methodologies in areas such as Automatic Fare Collection (AFC), Resource planning and scheduling (MRP), Supply Chain Management (SCM), and Production Floor Automation, for customers around the globe.

He graduated from the University of Colombo, Sri Lanka, with a degree in computer science in 2006. He lives in Perth, Australia, and is currently working full-time alongside Nathan Rozetals on the development of a next-generation automatic fare collection product.

Martin Jurča works as a tech leader and developer for Seznam.cz. He is passionate about TypeScript, frontend technologies, and pushing the boundaries of what is possible. Martin gave a speech at JSConf Iceland, is a regular speaker at tech conferences in Brno, and is a regular tutor at the Seznam IT academy.

Martin likes to relax with a good book or on a hike, or with his friends over a bottle of wine or a cup of tea. He lives in Brno with his wife and baby daughter.

I would like to thank my amazing wife for all the support she gives me and for every day I get to spend with her. I would also like to thank my daughter for motivating me to be a better person every day.

Packt is searching for authors like you

If you're interested in becoming an author for Packt, please visit `authors.packtpub.com` and apply today. We have worked with thousands of developers and tech professionals, just like you, to help them share their insight with the global tech community. You can make a general application, apply for a specific hot topic that we are recruiting an author for, or submit your own idea.

Table of Contents

Section 4: Extending the Capabilities of RESTful Web Services

Preface

This book is about hands-on development of TypeScript 3 APIs for beginners, intermediate, and expert-level software developers. The purpose of this book is to increase your practical knowledge by going through practical uses of features of TypeScript 3, OpenAPI design, database communication, Continuous Integration and Continuous Delivery practices, and also GraphQL fundamentals.

It is common to start creating APIs without meaning, or APIs that are not extensible at all. This book starts from scratch by looking at how to design and create a complete business capability through an API, and also provide an entire ecosystem that allows the API to keep evolving all the time, thus generating value more quickly.

All of the chapters in this book provide meaningful example applications for the topics covered, enabling you to put the practices demonstrated to use in real-life projects.

Who this book is for

This book is for anyone interested in developing APIs using TypeScript 3, applying best practices such as TDD, the OpenAPI Specification, Continuous Integration, Continuous Delivery, and GraphQL. You will benefit from prior experience with, and a basic understanding of JavaScript, but it is not mandatory. Even beginners will benefit a lot from the content of this book, since it focuses on examples that will help you when putting your ideas into production.

What this book covers

Chapter 1, *Introduction to RESTful API Development*, explains RESTful concepts in detail to help you to develop and run RESTful services. The main objective is to show comprehensive examples that transfer the concepts to real scenarios and help you to understand the definitions in a straightforward way.

Chapter 2, *Principles of Designing RESTful APIs*, prepares you for getting familiar with best practices. Having learned how to simplify operations, how to organize endpoints, how to name objects, and coding standardization in the beginning, this chapter will help you to understand the planning part easily.

Chapter 3, *Designing RESTful APIs with OpenAPI and Swagger*, focuses on core principles for creating web services. Instead of coding from the start, it describes how to design a web service first, then make it ready to code. Also, this chapter describes OpenAPI principles and implementation principles that can help readers to design their web services to support future changes and requirements.

Chapter 4, *Setting Up Your Development Environment*, covers the development environment—one of the key elements for most developers. Most developers get frustrated with configurations and tooling. You will learn how to set up a Node.js-based web server to serve your web service. You will also learn transpiling routines to transpile your TypeScript code to JavaScript. We will also cover Linters, which define semantic coding standards and check the source code while coding.

Chapter 5, *Building Your First API, Hello World*, mainly focuses on how to start the app that will serve the web service. The chapter describes file organization and folder structures for a more maintainable and scalable code base. Then, it focuses on how to define routes with a classic *Hello World* output as a result of a web service call. Finally, it shows the controller logic that will run when a certain endpoint is called.

Chapter 6, *Handling Requests and Responses*, covers the steps to take after creating the first route, that is, determining which properties you need while handling the requests that you receive, and also creating other routes. With that being said, it is really important to return meaningful responses in order to change/update application states. This chapter also covers methodologies that will be helpful in testing the application, such as not directly using request/response parameters in methods.

Chapter 7, *Formatting the APIs – Output*, introduces content negotiation, output formats, and the HAL JSON format to explain stateless API conventions. In the *Data serialization* section, you will learn how to convert resource objects to JSON objects and JSON arrays. Since JSON is standard nowadays, we mainly focus on that format. We also talk about how to expose data as XML.

Chapter 8, *Working with Databases and ODMs*, addresses a key point—persisting data for every web service. At any given time, a web service should return/serve the same data to simultaneous API calls. This chapter introduces setting up a MongoDB server and connecting it to your API. We will create some simple database wrapper methods that will help our internal logic while not mixing it with an external dependency.

Chapter 9, *Securing Your API*, discusses authorization techniques and authenticating users by using JWT-based tokens or basic authentication. Moving on, we use tools such as Passport and also look at security best practices. The chapter then describes the importance of serving API with SSL, and finally, it teaches you how to validate data to avoid exposing sensitive information.

Chapter 10, *Error Handling and Logging,* focuses on how to handle errors, starting with how to catch them and how to describe what an error is and is not. Without meaningful error messages, errors are hard to debug. Error messages should only describe the error itself and should not expose any sensitive data inside the error. This chapter explains how to write understandable error messages with minimum information.

Chapter 11, *Creating a CI/CD Pipeline for Your API,* covers DevOps—a must for almost every application life cycle. The containerization of an environment, using Continuous Integration services, running tests before deployment, and getting build notifications are the focus of this chapter. You'll learn how to create a pipeline by Dockerizing your environment using Travis CI, Google Cloud Platform, and GitHub.

Chapter 12, *Developing RESTful APIs with Microservices,* covers microservices—a hot topic nowadays. When you have your own RESTful API, you will start to think about how to run each service independently and how to control them inside their own medium. This chapter starts with a definition of what a microservice is and is not. Then, we continue with the isolation of APIs within an environment so that they can run autonomously. We also explore the possibilities for splitting an existing API into a smaller, more scalable microservice.

Chapter 13, *Flexible APIs with GraphQL,* looks at GraphQL, a new approach to serving data. Some people even define GraphQL as REST 2.0. We take a look at the differences and similarities between GraphQL and REST. You will learn how to add support for GraphQL to your existing RESTful API. With examples on querying data, and validating and executing a query, readers will learn how to reap the benefits of GraphQL for their RESTful APIs.

To get the most out of this book

1. You need knowledge of the following:

   - The JavaSript programming language
   - TypeScript basics
   - Web application concepts

- NoSQL database concepts
- Cloud computing basics

2. The following tools will be used in the book:

- TypeScript 3
- MongoDB
- Travis CI
- Swagger
- Robo 3T (Formerly Robomongo)
- Visual Studio Code
- Google Cloud Platform
- GitHub

Download the example code files

You can download the example code files for this book from your account at `www.packt.com`. If you purchased this book elsewhere, you can visit `www.packt.com/support` and register to have the files emailed directly to you.

You can download the code files by following these steps:

1. Log in or register at `www.packt.com`.
2. Select the **SUPPORT** tab.
3. Click on **Code Downloads & Errata**.
4. Enter the name of the book in the **Search** box and follow the onscreen instructions.

Once the file is downloaded, please make sure that you unzip or extract the folder using the latest version of:

- WinRAR/7-Zip for Windows
- Zipeg/iZip/UnRarX for Mac
- 7-Zip/PeaZip for Linux

The code bundle for the book is also hosted on GitHub at `https://github.com/PacktPublishing/Hands-On-RESTful-Web-Services-with-TypeScript-3`. In case there's an update to the code, it will be updated on the existing GitHub repository.

We also have other code bundles from our rich catalog of books and videos available at https://github.com/PacktPublishing/. Check them out!

Download the color images

We also provide a PDF file that has color images of the screenshots/diagrams used in this book. You can download it here: https://www.packtpub.com/sites/default/files/downloads/9781789956276_ColorImages.pdf.

Conventions used

There are a number of text conventions used throughout this book.

CodeInText: Indicates code words in the text, database table names, folder names, filenames, file extensions, pathnames, dummy URLs, user input, and Twitter handles. Here is an example: "First, go to the application folder, that is, my-first-express-app, and install the dependencies."

A block of code is set as follows:

```
if (isNaN(port)) {
  // named pipe
  return val;
}

if (port >= 0) {
  // port number
  return port;
}
```

When we wish to draw your attention to a particular part of a code block, the relevant lines or items are set in bold:

```
"name": "testing-nodejs-and-npm",
  "version": "1.0.0",
  "description": "",
  "main": "npm-app.js",
  "scripts": {
    "start": "node app.js"
  },
```

Any command-line input or output is written as follows:

```
$ node -v
```

Bold: Indicates a new term, an important word, or words that you see onscreen. For example, words in menus or dialog boxes appear in the text like this. Here is an example: "Search for **Cloud Endpoints** and click on **GO TO CLOUD ENDPOINTS**."

 Warnings or important notes appear like this.

 Tips and tricks appear like this.

Get in touch

Feedback from our readers is always welcome.

General feedback: If you have questions about any aspect of this book, mention the book title in the subject of your message and email us at customercare@packtpub.com.

Errata: Although we have taken every care to ensure the accuracy of our content, mistakes do happen. If you have found a mistake in this book, we would be grateful if you would report this to us. Please visit www.packt.com/submit-errata, selecting your book, clicking on the Errata Submission Form link, and entering the details.

Piracy: If you come across any illegal copies of our works in any form on the Internet, we would be grateful if you would provide us with the location address or website name. Please contact us at copyright@packt.com with a link to the material.

If you are interested in becoming an author: If there is a topic that you have expertise in and you are interested in either writing or contributing to a book, please visit authors.packtpub.com.

Reviews

Please leave a review. Once you have read and used this book, why not leave a review on the site that you purchased it from? Potential readers can then see and use your unbiased opinion to make purchase decisions, we at Packt can understand what you think about our products, and our authors can see your feedback on their book. Thank you!

For more information about Packt, please visit packt.com.

Section 1: Unraveling API Design

In this section, you will gain an understanding of the inner workings of API design. You will also learn best practices for creating APIs and how to plan your APIs for future needs.

The following chapters are included in this section:

- Chapter 1, *Introduction to RESTful Web Service Development*
- Chapter 2, *Principles of Designing RESTful APIs*
- Chapter 3, *Designing RESTful APIs with OpenAPI and Swagger*

Introduction to RESTful API Development 1

In order to get the best from any technology, it is necessary to understand deeply what the technology does. Throughout this chapter, RESTful concepts will be explained in detail with necessary elements to help you to develop and run RESTful services. The main objective of the chapter is to show you comprehensive examples that transpose these concepts to real-world scenarios and help you to understand their definitions in an easier way.

Currently, there is a huge necessity for applications to scale when necessary; they need to be fast, secure, portable, and trustable. These applications no longer serve one specific purpose; in fact, they work to solve corporative issues and meet the needs of end users.

APIs are the core elements for creating web and mobile applications. They provide structured data for your applications. This chapter mainly focuses on the definition of **Representational State Transfer (REST)** and RESTful principles, and describes the primary logic behind these architectural principles as well as the standards to follow while creating a web service.

The following topics will be covered in this chapter:

- The REST definition
- RESTful styles
- The general concepts regarding REST
- HTTP methods and verbs

Technical requirements

There are no requirements for this chapter; this chapter will lay the foundation for what this book is going to cover.

What is REST?

RESTful web services are services built according to REST principles. The idea is to have them designed to essentially work well on the web. Alright, but what is REST? Let's start from the beginning, that is, by defining REST.

The REST style is a set of software engineering practices that contains constraints that should be used in order to create web services in distributed hypermedia systems. REST is not a tool and neither is it a language; in fact, REST is agnostic of protocols, components, and languages.

It is important to say that REST is an architectural style and not a toolkit. REST provides a set of design rules in order to create stateless services that are shown as resources and, in some cases, sources of specific information such as data and functionality. The identification of each resource is performed by its unique **Uniform Resource Identifier (URI)**. REST describes simple interfaces that transmit data over a standardized interface such as HTTP and HTTPS without any additional messaging layer, such as **Simple Object Access Protocol (SOAP)**.

The consumer will access REST resources via a URI using HTTP methods (this will be explained in more detail later). After the request, it is expected that a representation of the requested resource is returned. The representation of any resource is, in general, a document that reflects the current or intended state of the requested resource.

REST architectural styles

The REST architectural style describes six constraints. These constraints were originally described by Roy Fielding in his Ph.D. thesis (https://www.ics.uci.edu/~fielding/pubs/dissertation/rest_arch_style.htm). They include the following:

- Uniform interface
- Stateless
- Cacheable
- Client-server architecture
- A layered system
- Code on demand (optional)

We will discuss them all minutely in the following subsections.

Uniform interface

Uniform interface is a constraint that describes a contract between clients and servers. One of the reasons to create an interface between them is to allow each part to evolve regardless of each other. Once there is a contract aligned with the client and server parts, they can start their works independently because, at the end of the day, the way that they will communicate is firmly based on the interface:

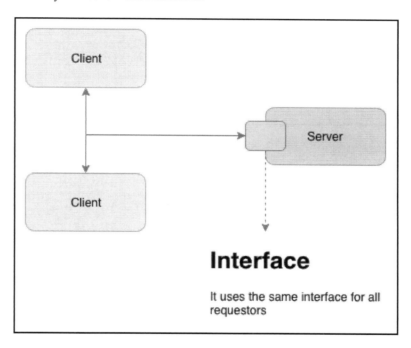

The uniform interface is divided into four main groups, called **principles**:

- **Resource-based**
- **The manipulation of resources using representations**
- **Self-descriptive messages**
- **Hypermedia as the Engine of Application State (HATEOAS)**

Let's talk more about them.

Resource-based

One of the key things when a resource is being modeled is the URI definition. The URI is what defines a resource as unique. This representation is what will be returned for clients. If you decided to perform GET to the offer URI, the resource that returns should be a resource representing an order containing the ID order, creation date, and so on. The representation should be in JSON or XML.

Here is a JSON example:

```
{
    id : 1234,
    creation-date : "1937-01-01T12:00:27.87+00:20",
    any-other-json-fields...
}
```

Here is an XML example:

```
<order>
    <id>1234</id>
    <creation-date>1937-01-01T12:00:27.87+00:20</creation-date>
    any-other-xml-fields
</order>
```

The manipulation of resources using representations

Following the happy path, when the client makes a request to the server, the server responds with a resource that represents the current state of its resource. This resource can be manipulated by the client. The client can request what kind it desires for the representation such as JSON, XML, or plain text.

When the client needs to specify the representation, the HTTP Accept header is used.

Here you can see an example in plain text:

```
GET https://<HOST>/orders/12345
Accept: text/plain
```

The next one is in JSON format:

```
GET https://<HOST>/orders/12345
Accept: application/json
```

Self-descriptive messages

In general, the information provided by the RESTful service contains all the information about the resource that the client should be aware of. There is also a possibility of including more information than the resource itself. This information can be included as a link. In HTTP, it is used as the content-type header and the agreement needs to be bilateral—that is, the requestor needs to state the media type that it's waiting for and the receiver must agree about what the media type refers to.

Some examples of media types are listed in the following table:

Extension	Document Type	MIME type
.aac	**AAC audio file**	audio/aac
.arc	**Archive document**	application/octet-stream
.avi	**Audio Video Interleave (AVI)**	video/x-msvideo
.css	**Cascading Style Sheets (CSS)**	text/css
.csv	**Comma-separated values (CSV)**	text/csv
.doc	**Microsoft Word**	application/msword
.epub	**Electronic publication (EPUB)**	application/epub+zip
.gif	**Graphics Interchange Format (GIF)**	image/gif
.html	**HyperText Markup Language (HTML)**	text/html
.ico	**Icon format**	image/x-icon
.ics	**iCalendar format**	text/calendar
.jar	**Java Archive (JAR)**	application/java-archive
.jpeg	**JPEG images**	image/jpeg
.js	**JavaScript (ECMAScript)**	application/javascript
.json	**JSON format**	application/json
.mpeg	**MPEG video**	video/mpeg
.mpkg	**Apple Installer Package**	application/vnd.apple.installer+xml

`.odt`	**OpenDocument text document**	`application/vnd.oasis.opendocument.text`
`.oga`	**OGG audio**	`audio/ogg`
`.ogv`	**OGG video**	`video/ogg`
`.ogx`	**OGG**	`application/ogg`
`.otf`	**OpenType font**	`font/otf`
`.png`	**Portable Network Graphics**	`image/png`
`.pdf`	**Adobe Portable Document Format (PDF)**	`application/pdf`
`.ppt`	**Microsoft PowerPoint**	`application/vnd.ms-powerpoint`
`.rar`	**RAR archive**	`application/x-rar-compressed`
`.rtf`	**Rich Text Format (RTF)**	`application/rtf`
`.sh`	**Bourne shell script**	`application/x-sh`
`.svg`	**Scalable Vector Graphics (SVG)**	`image/svg+xml`
`.tar`	**Tape Archive (TAR)**	`application/x-tar`
`.ts`	**TypeScript file**	`application/typescript`
`.ttf`	**TrueType Font**	`font/ttf`
`.vsd`	**Microsoft Visio**	`application/vnd.visio`
`.wav`	**Waveform Audio Format**	`audio/x-wav`
`.zip`	**ZIP archive**	`application/zip`
`.7z`	**7-zip archive**	`application/x-7z-compressed`

There is also a possibility of creating custom media types. A complete list can be found at `https://developer.mozilla.org/en-US/docs/Web/HTTP/Basics_of_HTTP/MIME_types/Complete_list_of_MIME_types`.

HATEOAS

HATEOAS is a way that the client can interact with the response by navigating within it through the hierarchy in order to get complementary information.

For example, here the client makes a GET call to the `order` URI :

```
GET https://<HOST>/orders/1234
```

The response comes with a navigation link to the items within the 1234 order, as in the following code block:

```
{
    id : 1234,
    any-other-json-fields...,
    links": [
        {
            "href": "1234/items",
            "rel": "items",
            "type" : "GET"
        }
    ]
}
```

What happens here is that the link fields allow the client to navigate until 1234/items in order to see all the items that belong to the 1234 order.

Stateless

Essentially, stateless means that the necessary state during the request is contained within the request and it is not persisted in any hypothesis that could be recovered further. Basically, the URI is the unique identifier to the destination and the body contains the state or changeable state, or the resource. In other words, after the server handles the request, the state could change and it will send back to the requestor with the appropriate HTTP status code:

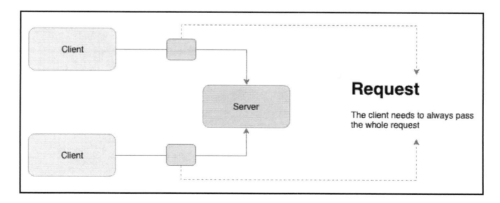

In comparison to the default session scope found in a lot of existing systems, the REST client must be the one that is responsible in providing all necessary information to the server, considering that the server should be idempotent.

Stateless allows high scalability since the server will not maintain sessions. Another interesting point to note is that the load balancer does not care about sessions at all in stateless systems.

In other words, the client needs to always pass the whole request in order to get the resource because the server is not allowed to hold any previous request state.

Cacheable

The aim of caching is to never have to generate the same response more than once. The key benefits of using this strategy are an increase in speed and a reduction in server processing.

Essentially, the request flows through a cache or a series of caches, such as local caching, proxy caching, or reverse proxy caching, in front of the service hosting the resource. If any of them match with any criteria during the request (for example, the timestamp or client ID), the data is returned based on the cache layer, and if the caches cannot satisfy the request, the request goes to the server:

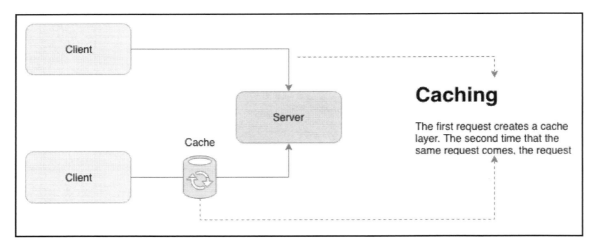

Client-server architecture

The REST style separates clients from a server. In short, whenever it is necessary to replace either the server or client side, things should flow naturally since there is no coupling between them. The client side should not care about data storage and the server side should not care about the interface at all:

A layered system

Each layer must work independently and interact only with the layers directly connected to it. This strategy allows passing the request without bypassing other layers. For instance, when scaling a service is desired, you might use a proxy working as a load balancer—that way, the incoming requests are deliverable to the appropriate server instance. That being the case, the client side does not need to understand how the server is going to work; it just makes requests to the same URI.

The cache is another example that behaves in another layer, and the client does not need to understand how it works either:

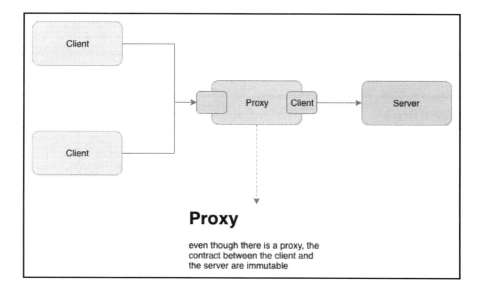

Code on demand

In summary, this optional pattern allows the client to download and execute code from the server on the client side. The constraint says that this strategy improves scalability since the code can execute independently of the server on the client side:

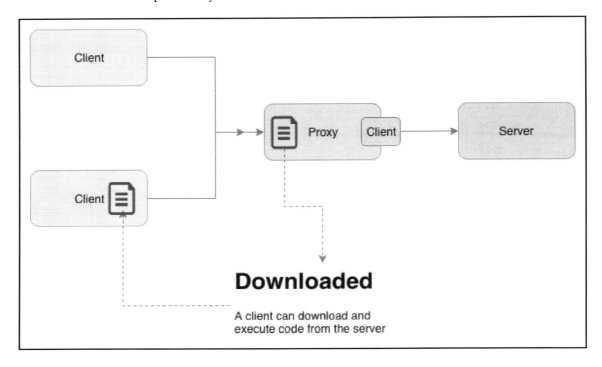

Downloaded

A client can download and execute code from the server

HTTP methods for RESTful services

The use of HTTP verbs allows a clear understanding of what an operation is going to do. In general, the primary or most commonly used HTTP verbs are POST, GET, PUT, PATCH, and DELETE, which stand for create, read, update (PATCH and PUT), and delete, respectively. Of course, there are also a lot of other verbs, but they are not used as frequently:

Method	HTTP method description
GET	GET is the most common HTTP verb. Its function is to retrieve data from a server at the specified resource. For example, a request made to the GET `https://<HOST>/customers` endpoint will retrieve all customers in a list format (if there is no pagination). There is also the possibility of retrieving a specific customer such as GET `https://<HOST>/customers/1234`; in this instance, only the customer with the `1234` ID will be retrieved. It is important to add that the GET request only retrieves data and does not modify any resources; it's considered a safe and idempotent method.
HEAD	The HEAD method does exactly what GET does, except that the server replies without the body.
POST	The most common usage of POST methods is to create a new resource.
PATCH	The PATCH method is used to apply partial modifications to a resource, such as updating a name or a date, but not the whole resource.
PUT	Different from the PATCH method, the PUT method replaces all of the resource.
DELETE	The DELETE method removes a resource.
CONNECT	The CONNECT method converts the connection request to a transparent TCP/IP tunnel, generally to facilitate encrypted communication with SSL (HTTPS) through an unencrypted HTTP proxy.
OPTIONS	The OPTIONS method returns the HTTP methods supported by the server for the specified URL.
TRACE	The TRACE method returns the same request that is sent to see whether there were changes and/or additions made by intermediate servers.

In order to explain these methods in detail, let's consider a simple entity called **Customer**. We will imagine that there is an API called Customers available through HTTP and its destination is a NoSQL database like in the following diagram:

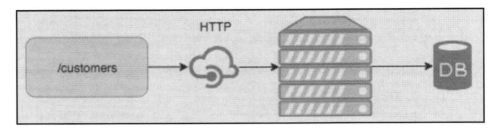

Considering that the database is empty, what should happen if we call the GET method by pointing to GET /CUSTOMERS? If you think that it should retrieve nothing, then you are absolutely right! Here is a request example:

```
GET https://<HOST>/customers
```

And here is the response:

```
[]
```

So, considering that there is no customer yet, let's create one using the POST method:

Request endpoint	Sample Request	Status code response

POST /customers	`{`
	`    "first_name":"John",`
	`    "last_name":"Doe"`
	`}`
		HTTP status code **201**

Considering the POST method is successful, the HTTP status code should be **201** (we will talk more about status codes in Chapter 2, *Principles of Designing RESTful APIs*) and calling the GET method now; the response should be similar to the following code block (we're considering that there is just one single consumer of this endpoint and we are that consumer, so we can guarantee the data behavior into the server):

Request endpoint	Sample Request	Status code response

GET /customers	`[`
	`  {`
	`    "id" : 1,`
	`    "first_name":"John",`
	`    "last_name":"Doe"`
	`  }`
	`]`
		The HTTP status code is **200**.

What if we add two more customers and call the GET method again? Take a look at the following code block:

Request endpoint	Sample Response	Status code response
GET /customers	```[{ "id" : 1, "first_name" : "John", "last_name" : "Doe" }, { "id" : 2, "first_name" : "Mary", "last_name" : "Jane" }, { "id" : 3, "first_name" : "Jane", "last_name" : "Doe" }]```	The HTTP status code is **200**.

Nice, so if we call the GET method for a specific ID, the response should be like the following:

Request endpoint	Sample Response	Status code response
GET /customers/1	```{ "id" : 1, "first_name":"John", "last_name":"Doe" }```	The HTTP status code is **200**.

We can also call POST methods for parent resources, such as /customers/1/orders, which will create new resources under the customer with ID 1. The same applies to the GET method, but only to retrieve the data as mentioned previously.

 Remember that the GET method should never modify any resource.

Okay, now that we know how to create and retrieve the resources, what if we want to change any information, such as the last name for `John Doe`? This one is easy—we just have to call the `PATCH` method in the same way that we called the `POST` method:

Request endpoint	Sample Request	Status code response

`PATCH /customers/1`	`{`
	`   "last_name" : "Doe Jr."`
	`}`
		The HTTP status code is **204**.

We can also change the information in a parent using a path as follows:

Request endpoint	Sample Request	Status code response

`PATCH /customers/1/orders`	`{`
	`   "status" : "closed"`
	`}`
		The HTTP status code **204**.

And what if we need to change the whole resource? In that case, we should use the `PUT` method:

Request endpoint	Sample Request	Status code response

`PUT /customers/1`	`{`
	`   "first_name" : "Wolfgang",`
	`   "last_name" : "Amadeus Mozart"`
	`}`
		The HTTP status code is **201**.

Instead of `PATCH`, you can use the `PUT` method. To do so, all the parameters are required to be passed as the body into the request, even those parameters that haven't changed.

Finally, if you want to remove a resource, use the `DELETE` method:

Request endpoint	Status code response
`PUT /customers/1`	The HTTP status code is **204**.

Summary

In this chapter, we talked about the various HTTP methods and when each one of them should be used. We also discussed the principles of RESTful services. All elements presented in this chapter and the upcoming chapters will be applied in the case study.

We also covered the general idea of HTTP methods and verbs based on examples with JSON. The REST architectural style was described based on six constraints.

In the next chapter, we will present the best practices related to RESTful design such as API endpoint organization, different ways to expose an API service, how to handle large datasets, naming conventions, the HTTP status codes, and API versioning.

Questions

1. What method should be used in order to change the resource state?
2. What is the best way to create a new resource?
3. What are the REST architectural style constraints?
4. What is the only optional REST architectural style constraint?
5. If you need to change only two fields in a resource that contains more than two fields, what verb should you use?
6. What is the verb used to update the whole resource?
7. What does GET do?

Further reading

In order to improve your knowledge of REST, RESTful, and RESTful styles, the following books are recommended to be read, as they will be helpful in the coming chapters:

- *Building RESTful Web Services with Java EE 8* (https://www.packtpub.com/application-development/building-restful-web-services-java-ee-8)
- *RESTful Web API Design with Node.js 10 – Third Edition* (https://www.packtpub.com/web-development/restful-web-api-design-nodejs-10-third-edition)
- *Building RESTful Web Services with Spring 5 – Second Edition* (https://www.packtpub.com/application-development/building-restful-web-services-spring-5-second-edition)

2
Principles of Designing RESTful APIs

Since we described what REST is in the previous chapter, we will now go over best practices for REST and RESTful APIs. In the upcoming chapters, we are going to create our own web services. Since we have already learned how to simplify operations, how to organize endpoints, and how to name and perform code standardization, this chapter will help you understand the planning part easily. Before you start coding, we will give you a brief overview of what type of decisions you need to make while creating APIs.

The following topics will be covered in this chapter:

- Organizing API endpoints
- Different ways to expose an API service
- Working with large datasets
- Naming conventions
- HTTP status codes
- API versioning

Technical requirements

This chapter continues from the setup completed in the previous chapter and does not have any additional requirements.

Organizing API endpoints

The first important thing that you need to know about organizing API endpoints is that a resource is any information that can be named, which means that a URI should be an entity-based order and not datatype-based current-order-items.

It is good practice to separate internal/private APIs from external/public APIs, for example, by keeping private APIs in the same VPC or making them available only to the company or a specific domain, and making public ones available to any consumer, and sometimes over the internet. It is really common to expose business capabilities externally and foundation capabilities internally. It is important to organize and expose APIs properly, since they will always be used—and they will be used properly. Taking advantage of the business capabilities and creating business-oriented APIs allows an organization to get the most value out of those APIs.

There are three categories of APIs, which allows organizations to create business-oriented APIs:

- Resource-based APIs
- Experience-based APIs
- Capability-based APIs

We will discuss these in more detail in the following sections.

Resource-based APIs

This category essentially states that each resource represents a specific collection of information regarding an entity or collection of entities of a given type. Resources are organized based on the information that's provided by the domain itself. In general, APIs use the REST style to perform operations that reflect the state of a given resource. From an enterprise perspective, exposed resources based on APIs allow for mirroring key business concepts, rather than silos.

Here, you can see examples where the resource is the user and its representation is related to login and logout actions:

```
/users/login
/users/logout
```

The following example shows information such as ratings, titles, and recommendations under a specific user ID. The same idea applies to items from a given order ID:

```
/users/<id>/ratings/title
/users/<id>/recommendations

/stores/orders/<id>/items
```

Experience-based APIs

Experience-based APIs represent a different approach to resource-based APIs. Here, the concept is **One-Size-Fits-All (OSFA)**. Basically, the key to experience-based APIs is providing endpoints that are focused on the experience itself, which isn't only the user's experience—it could be the home screen experience.

The following example shows the `dashboard-screen` operation being directly connected to the Nintendo resource, thus expressing the experience of the screen on Nintendo:

```
/nintendo/dashboard-screen
```

This concept came from Netflix and, in general, it is an extension of the proxy pattern, where new proxy endpoints will be designed, developed, and configured to focus on the business requirements of the API consumers.

The following diagram shows how the **Experience-based APIs** building blocks connect to the external world (web, mobile, and so on) and with downstream systems such as the **Legacy API**:

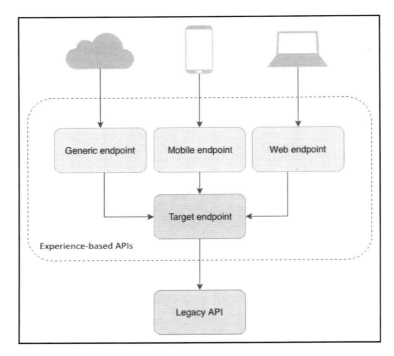

Capability-based APIs

An interesting way to expose APIs is through capabilities, which is a common strategy that's adopted by huge enterprises like B2B. Enterprises often adopt business architecture to develop strategies that converge to align across technology and business. One of the most common techniques that's used in business architecture is business capability modeling. This technique can be used to represent an organization's business anchor model. Such representations are independent of structures or people.

It is quite easy for organizations to leverage business capability modeling as a way to express and explore what the organization does. Based on that, people can make decisions regarding planning. These accomplishments are intrinsic and related to business leaders. They have to work closely, in that they have to cooperate and collaborate, considering that this strategy should be driven by business units, rather than IT.

An example of this would be a company that wants to expose their order process as an API, like so:

```
[GET] /stores/products/<id>/items --> for the store to find items
[GET] /procuts/<id>/items--> the quantity of an item among the stores
```

Working with large datasets

Everything goes smoothly when a dataset is small, even though there is no fixed value definition of what small means. Being agnostic, what happens if a dataset is really big and, for instance, a user decides to perform a GET request on an endpoint? Should we retrieve all resources at the same time? The answer is no, and there are a few possibilities for handling such a situation. In this section, we will take a look at some of the ways to handle these kinds of situations.

Pagination

Let's treat this issue like a real scenario. Considering that there is an endpoint that retrieves all orders from a retail company, such as GET /orders, the most obvious idea is to use a pagination strategy to allow the requester to define how many records they want to get. Thus, the requester has to send a query parameter telling the server how many items should be retrieved, given the limit and offset parameters. The offset and limit query parameters are very popular and are being adopted as a pattern for pagination. Basically, the offset will tell the service to start, based on the value passed as a parameter—the limit will tell you how long it will run for.

The following URI shows an example of calling an endpoint with pagination in place:

```
GET /items?offset=100&limit=20
```

This request will retrieve all items, starting from item number 100 until 120. At the moment, there is no sort strategy, so the result will be simply the first 20 items after 100:

```
[
    {
        resource_100
    },
    ...
    {
        resource_119
    }
]
```

There is also another possibility—using keyset or pagination links as the continuation point. It often uses timestamps as a key. For example, take the `GET /items?limit=20` request. The keyset values are as follows:

- `first`: The first item based on the sort strategy chosen
- `previous`: The previous item based on the sort strategy chosen
- `self`: The item itself
- `next`: The next item based on the sort strategy chosen
- `last`: The last item based on the sort strategy chosen

The following example shows how to use paging information with the next and previous buckets:

```
{
    "data" : [
        {
            resource 1
        },
        {
            resource 2
        },
        ...
        {
            resource 20
        }
    ],
    "paging": {
        "previous": "https://{URI}/items?since=TIMESTAMP1"
        "next": "https://{URI}/items?since=TIMESTAMP2"
    }
}
```

We could also use the `since` query parameter based on the timestamp to get the previous and next batch of data based on the response of the first request. If we took the preceding response as an example and hit `next`, we'd get data starting on the last result on the first request we made, which means `21` and so on. Of course, we can combine the `limit` query parameter as well, for example, `GET /items?since={TIMESTAMP}&limit=20`.

Sorting

Like limiting the response, sorting is a good strategy to leverage when API design is being done. In general, what the end user will see is just a new query parameter called `sort`, which tells the service which fields should be used for the sorting strategy. Also, the URI may contain information about the sorting strategy, such as ascending, descending, and so on.

The value of the `sort` parameter is a comma-separated list of sort keys, like a map. Likewise, sort directions can be appended to each sort key pair, separated by the : character:

```
https://<URI>/items/sort=name,description,size
```

In this example, the services will sort everything based on the first key, which is `name`; then, the services will sort by the `description` field, and finally, by the `size` field. The supported sort directions are either `asc` for ascending or `desc` for descending:

```
https://<URI>/items?sort=name:asc,description,size
https://<URI>/items?sort=name:asc,description:desc,size
https://<URI>/items?sort=name:asc,description:asc,size
https://<URI>/items?sort=name,description:asc,size
https://<URI>/items?sort=name:asc,description,size:desc
```

A sort direction may be (optionally) given by the caller for each key. However, if it's not provided, a default will be set by the server, which is the one that's implemented by the backend service if nothing is provided through the URI.

Filtering

Sometimes, a resource contains more data than is needed by the requester. In this case, the API should have the ability to filter elements in or out from the resource.

> Filter `in` or `out` doesn't mean that the resource will change. Remember that `GET` methods are idempotent. It will only remove from the response, but the original information will still be available.

Filtering can be implemented as a query parameter that's named for the field to be filtered on. The value needs to be the value it needs to filter for. You can refer to the definition of filtering at https://jsonapi.org/recommendations/#filtering.

For example, the following is a request for all items associated with a particular order with a name of coffee:

```
GET http://<HOST>/order/<id>/items?filter[name]=coffee
```

Also, multiple filter values could be combined in a comma-separated list:

```
GET http://<HOST>/order/<id>/items?filter[name]=coffee,milk
```

Furthermore, multiple filters can also be applied to the same request as another filter:

```
GET
http://<HOST>/order/<id>/items?filter[name]=coffee,milk&filter[category]=or
ganic
```

There is also the possibility to include the following operators:

- in: Getting items within the range between 5 and 20:

  ```
  GET http://<HOST>/order/<id>/items?filter[size]=in:5,20
  ```

- nin: Getting items that are out of a range:

  ```
  GET http://<HOST>/order/<id>/items?filter[size]=nin:5,20
  ```

- neq: Getting items not matching a specific value:

  ```
  GET http://<HOST>/order/<id>/items?filter[category]=neq:built-in
  ```

- gt: Getting items greater than a specific value:

  ```
  GET http://<HOST>/order/<id>/items?filter[size]=gt:5
  ```

- gte: Getting items greater than a specific value, inclusive:

  ```
  GET http://<HOST>/order/<id>/items?filter[size]=gte:5
  ```

- lt: Getting items lower than a specific value:

  ```
  GET http://<HOST>/order/<id>/items?filter[size]=lt:25
  ```

- lte: Getting items lower than a specific value, inclusive:

  ```
  GET http://<HOST>/order/<id>/items?filter[size]=lte:25
  ```

Naming conventions

One of the keys to achieving a good RESTful design is naming the HTTP verbs appropriately. It is really important to create understandable resources that allow people to easily discover and use your services. A good resource name implies that the resource is intuitive and clear to use. On the other hand, the usage of HTTP methods that are incompatible with REST patterns creates noise and makes the developer's life harder. In this section, there will be some suggestions for creating clear and good resource URIs.

It is good practice to expose resources as nouns instead of verbs. Essentially, a resource represents a *thing*, and that is the reason you should use nouns. Verbs refer to actions, which are used to factor HTTP actions.

Three words that describe good resource naming conventions are as follows:

- **Understandability**: The resource's representation format should be understandable and utilizable by both the server and the client
- **Completeness**: A resource should be completely represented by the format
- **Linkability**: A resource can be linked to another resource

Some example resources are as follows:

```
Users of a system
Blogs posts
An article
Disciplines in which a student is enrolled
Students in which a professor teaches
A blog post draft
```

Each resource that's exposed by any service in a best-case scenario should be exposed by a unique URI that identifies it. It is quite common to see the same resource being exposed by more than one URI, which is definitely not good. It is also good practice to do this when the URI makes sense and describes the resource itself clearly. URIs need to be predictable, which means that they have to be consistent in terms of data structure. In general, this is not a REST required rule, but it enhances the service and/or the API.

A good way to write good RESTful APIs is by writing them while having your consumers in mind. There is no reason to write an API and name it while thinking about the APIs developers rather than its consumers, who will be the people who are actually consuming your resources and API (as the name suggests). Even though the resource now has a good name, which means that it is easier to understand, it is still difficult to understand its boundaries. Imagine that services are not well named; bad naming creates a lot of chaos, such as business rule duplications, bad API usage, and so on.

In addition to this, we will explain naming conventions based on a hypothetical scenario.

Let's imagine that there is a company that manages orders, offers, products, items, customers, and so on.

Considering everything that we've said about resources, if we decided to expose a customer resource and we want to insert a new customer, the URI might be as follows:

```
POST https://<HOST>/customers
```

The hypothetical request body might be as follows:

```
{
    "fist-name" : "john",
    "last-name" : "doe",
    "e-mail"    : "john.doe@email.com"
}
```

Imagine that the previous request will result in a customer ID of 445839 when it needs to recover the customer. The GET method could be called as follows:

```
GET https://<HOST>/customers/445839
```

The response will look something like this:

```
sample body response for customer #445839:

{
    "customer-id": 445839,
    "fist-name"  : "john",
    "last-name"  : "doe",
    "e-mail"     : "john.doe@email.com"
}
```

The same URI can be used for the PUT and DELETE operations, respectively:

```
PUT https://<HOST>/customers/445839
```

The PUT body request might be as follows:

```
{
 "last-name" : "lennon"
}
```

For the DELETE operation, the HTTP request to the URI will be as follows:

```
DELETE https://<HOST>/customers/445839
```

Moving on, based on the naming conventions, the product URI might be as follows:

```
POST https://<HOST>/products
sample body request:

{
    "name"          :  "notebook",
    "description"   :  "and fruit brand"
}

GET https://<HOST>/products/9384

PUT https://<HOST>/products/9384

sample body request:

{
    "name" : "desktop"
}

DELETE https://<HOST>/products/9384
```

Now, the next step is to expose the URI for order creation. Before we continue, we should go over the various ways to expose the URI. The first option is to do the following:

```
POST https://<HOST>/orders
```

However, this could be outside the context of the desired customer. The order exists without a customer, which is quite odd. The second option is to expose the order inside a customer, like so:

```
POST https://<HOST>/customers/445839/orders
```

Based on that model, all orders belong to user 445839. If we want to retrieve those orders, we can make a GET request, like so:

```
GET https://<HOST>/customers/445839/orders
```

As we mentioned previously, it is also possible to write hierarchical concepts when there is a relationship between resources or entities. Following the same idea of orders, how should we represent the URI to describe items within an order and an order that belongs to user 445839?

First, if we would like to get a specific order, such as order 7384, we can do that like so:

```
GET https://<HOST>/customers/445839/orders/7384
```

Following the same approach, to get the items, we could use the following code:

```
GET https://<HOST>/customers/445839/orders/7384/items
```

The same concept applies to the `create` process, where the URI is still the same, but the HTTP method is `POST` instead of `GET`. In this scenario, the body also has to be sent:

```
POST https://<HOST>/customers/445839/orders/7384

{
    "id" : 7834,
    "quantity" : 10
}
```

Now, you should have a good idea of what the `GET` operation offers in regard to orders. The same approach can also be applied so that you can go deeper and get a specific item from a specific order and from a specific user:

```
GET https://<HOST>/customers/445839/orders/7384/items/1
```

Of course, this hierarchy applies to the `PUT`, `PATCH`, and `POST` methods, and in some cases, the `DELETE` method as well. It will depend on your business rules; for example, can the item be deleted? Can I update an order?

Versioning

As APIs are being developed, gathering more business rules for their context on a day-to-day basis, generating tech debits and maturing, there often comes a point where teams need to release breaking functionality. It is also a challenge to keep their existing consumers working perfectly. One way to keep them working is by versioning APIs.

Breaking changes can get messy. When something changes abruptly, it often generates issues for consumers, as this usually isn't planned and directly affects the ability to deliver new business experiences.

There is a variant that says that APIs should be versionless. This means that building APIs that won't change their contract forces every change to be viewed through the lens of backward compatibility. This drives us to create better API interfaces, not only to solve any current issues, but to allow us to build APIs based on foundational capabilities or business capabilities themselves. Here are a few tips that should help you out:

- **Put yourself in the consumer's shoes**: When it comes to product perspective, it is suggested that you think from the consumer's point of view when building APIs. Most breaking changes happen because developers build APIs without considering the consumers, which means that they are building something for themselves and not for the real users' needs.

- **Contract-first design**: The API interface has to be treated as a formal contract, which is harder to change and more important than the coding behind it. The key to API design success is understanding the consumer's needs and the business associated with it to create a reliable contract. This is essentially a good, productive conversation between the consumers and the producers.

- **Requires tolerant readers**: It is quite common to add new fields to a contract with time. Based on what we have learned so far, this could generate a breaking change. This sometimes occurs because, unfortunately, many consumers utilize a deserializer strategy, which is strict by default. This means that, in general, the plugin that's used to deserialize throws exceptions on fields that have never been seen before. It is not recommended to version APIs, but only because you need to add a new optional field to the contract. However, in the same way, we don't want to break changes on the client side. Some good advice is documenting any changes, stating that new fields might be added so that the consumers aren't surprised by any new changes.

- **Add an object wrapper**: This sounds obvious, but when teams release APIs without object wrappers, the APIs turn on hard APIs, which means that they are near impossible to evolve without having to make breaking changes. For instance, let's say your team has delivered an API based on JSON that returns a raw JSON array. So far, so good. However, as they continue, they find out that they have to deal with paging, or have to internationalize the service or any other context change. There is no way of making changes without breaking something because the return is based on raw JSON.

- **Always plan to version**: Don't think you have built the best turbo API in the world ever. APIs are built with a final date, even though you don't know it yet. It's always a good plan to build APIs while taking versioning into consideration.

Including the version in the URL

Including the version in the URL is an easy strategy for having the version number added at the end of the URI. Let's see how this is done:

```
https://api.domain.com/v1/
https://api.domain.com/v2/
https://api.domain.com/v3/
```

Basically, this model tells the consumers which API version they are using. Every breaking change increases the version number. One issue that may occur when the URI for a resource changes is that the resource may no longer be found with the old URI unless redirects are used.

Versioning in the subdomain

In regard to versioning in the URL, subdomain versioning puts the version within the URI but associated with the domain, like so:

```
https://v1.api.domain.com/
https://v2.api.domain.com/
https://v3.api.domain.com/
```

This is quite similar to versioning at the end of the URI. One of the advantages of using a subdomain strategy is that your API can be hosted on different servers.

Versioning on media types

Another approach to versioning is using MIME types to include the API version. In short, API producers register these MIME types on their backend and then the consumers need to include accept and content-type headers.

The following code lets you use an additional header:

```
GET https://<HOST>/orders/1325 HTTP/1.1
Accept: application/json
Version: 1

GET https://<HOST>/orders/1325 HTTP/1.1
Accept: application/json
Version: 2

GET https://<HOST>/orders/1325 HTTP/1.1
```

```
Accept: application/json
Version: 3
```

The following code lets you use an additional field in the accept/content-type header:

```
GET https://<HOST>/orders/1325 HTTP/1.1
Accept: application/json; version=1

GET https://<HOST>/orders/1325 HTTP/1.1
Accept: application/json; version=2

GET https://<HOST>/orders/1325 HTTP/1.1
Accept: application/json; version=3
```

The following code lets you use a Media type:

```
GET https://<HOST>/orders/1325 HTTP/1.1
Accept: application/vnd.<host>.orders.v1+json

GET https://<HOST>/orders/1325 HTTP/1.1
Accept: application/vnd.<host>.orders.v2+json

GET https://<HOST>/orders/1325 HTTP/1.1
Accept: application/vnd.<host>.orders.v3+json
```

Recommendation

When using a RESTful service, it is highly recommended that you use header-based versioning. However, the recommendation is to keep the version in the URL. This strategy allows the consumers to open the API in a browser, send it in an email, bookmark it, share it more easily, and so on. This format also enables human log readability.

There are also a few more recommendations regarding API versioning:

- **Use only the major version**: API consumers should only care about breaking changes.
- **Use a version number**: Keep things clear; numbering the API incrementally allows the consumer to track evolvability. Versioning APIs using timestamps or any other format only creates confusion in the consumer's mind. This also exposes more information about versioning than is necessary.
- **Require that the version has to be passed**: Even though this is more convenient from the API producer's perspective, starting with a version is a good strategy because the consumers will know that the API version might change and they will be prepared for that.

- **Document your API time-to-live policy**: Good documentation is a good path to follow. Keeping everything well-described will mean that consumers avoid finding out that there is no Version 1 available anymore because it has been deprecated. Policies allow consumers to be prepared for issues such as depreciation.

HTTP status codes

A huge list of well-defined HTTP status codes exists. This list of commonly used status codes is useful for designing RESTful APIs and leveraging the semantics that are defined for clear communication between producers and consumers.

It is highly recommended to not return a HTTP **200** status for all requests and encode the success or failure of the request in the response body. Keep in mind that there will be a specific HTTP status code for each scenario. Returning the proper status code will indicate the exact behavior that occurred on the server side, which allows the consumer to understand what happened there. There are five high-level HTTP status code classes, as described in the following table:

High-level status code	Class	Description
1xx	Informational	Continues the process after receiving the request.
2xx	Success	The action was received, understood, and accepted successfully.
3xx	Redirection	Further action required to complete the request.
4xx	Client Error	The request contains bad syntax or cannot be fulfilled. Possible reasons include authentication/authorization.
5xx	Server Error	The server failed to fulfil the request.

The following sections describe the most commonly used HTTP status codes in detail.

2xx – success

The variations of 2xx status codes, which represent success scenarios, can be compressed into four main 2xx status codes:

- **200 OK**: This is the most common happy path response, indicating that the request succeeded.
- **201 Created**: This indicates that a new resource has been successfully created. It is most commonly used as a response to POST request methods. This response should contain a location header, which provides the URL for the new resource that has been created. There is a recommendation to send the response body back with the newly created resource. Essentially, the response body is the same if the consumer hits the GET endpoint to retrieve the resource.
- **202 Accepted**: This is kind of equivalent to **201**. The only difference is that **202** is used for asynchronous operations, as it clearly indicates eventual consistency.
- **204 No Content**: This is also similar to **200**, but only when there is no response body. This is mostly used for DELETE requests.

3xx – redirection

The variations of 3xx redirection codes are as follows:

- **301 Moved Permanently**: Indicates that the resource has been relocated and that the new URL location should be provided on the location header.
- **304 Not Modified**: Indicates that the resource is the same as when the client last requested it. It indicates a safe request and is commonly used as part of a caching strategy. The client can send either an ETag or If-Modified-Since header, and the server can avoid sending the response body if the resource is already cached.

4xx – client error

This list shows the client error codes:

- **400 Bad Request**: The request could not be handled because of a syntax error. It could be a missing or required query parameter, unparseable JSON, a missing required JSON key, or a missing header on requests that require optimistic concurrency. It is good practice to include the reason for the error in the response body.

- **401 Unauthorized**: When the client does not provide any authentication mechanism, the server refuses to handle a request.
- **403 Forbidden**: When the server receives the request but the consumers are not allowed to use the method that's been requested, **403** should be retrieved. **403** only indicates permissions and not authentication. Any issue related to authentication is related to **401**, whereas **403** indicates that the client is unable to access the resource with the credentials that were provided.
- **404 Not Found**: One of the most famous status codes. **404** indicates that the resource does not exist.
- **405 Method Not Allowed**: A request has been made using a request method that is not compatible with this resource, for example, by using `GET` on a form, which requires data to be submitted via `POST` or `PUT`.
- **406 Not Acceptable**: Indicates that the request includes an `Accepts` header that the endpoint is unable to satisfy. This status code is only relevant if you support multiple response content types, though you may also encounter it during development with some particularly strict frameworks.
- **409 Conflict**: Most commonly used on optimistic concurrency on resource updates. 409 indicates that the current resource has changed since the last time it was requested.
- **415 Unsupported Media Type**: Indicates that the entity has a media type request that the server or resource does not support. For example, the consumer tries to post an image such as PNG and JSON, but the server is not prepared to process this format, since it expects a different one.
- **429 Too Many Requests**: Sometimes, it is necessary to handle requests based on a rate-limiting or throttling function. The response should include details that explain the condition regarding the too many requests limit.

5xx – server error

The variations of server error codes are as follows:

- **500 Internal Server Error**: This is a general server-side error without other information. Of course, it is not a good idea to just send **500** for any kind of server error. I am pretty sure that good errors and messages could be provided to help you troubleshoot the error.
- **503 Service Unavailable**: This is used to indicate a downstream service's dependency failed. It is quite similar to **502**.

Summary

In this chapter, we learned about best practices related to RESTful web services such as strategies for API endpoint organization. We also looked at different ways to expose an API service, how to handle large datasets, and designing APIs with pagination, filtering, and sorting in mind. Finally, we learned about naming conventions, the HTTP status codes list, and API versioning formats.

In the next chapter, we will look at how to design RESTful web services with OpenAPI and Swagger, focusing on the core principles while creating web services. Instead of coding from the start, we will describe how to design a web service first and prepare it for coding.

Questions

1. What does the `GET /users?offset=100&limit=20` operation do?
2. What should the URL look like if the client wants to sort by first name and last name and use a `GET` method to get the user's endpoint?
3. What are the three main items that describe a good resource naming convention?
4. If I want to create an endpoint to retrieve just one user, should the URI be in singular form?
5. What is the recommended strategy for API versioning?
6. What are the five HTTP status code classes?
7. If I call a `DELETE` method, which HTTP status code should be sent from the server side if `DELETE` succeeds?
8. Which HTTP status code should be sent when there are too many requests?

Further reading

To improve your knowledge on best practices regarding RESTful web services, the following books are recommended and will be helpful for the upcoming chapters:

- *Beginning API Development with Node.js* (https://www.packtpub.com/web-development/beginning-api-development-nodejs)
- *RESTful Web API Design with Node.js* (https://www.packtpub.com/web-development/restful-web-api-design-nodejs-10-third-edition)
- *RESTful Java Patterns and Best Practices* (https://www.packtpub.com/application-development/restful-java-patterns-and-best-practices)

3
Designing RESTful APIs with OpenAPI and Swagger

Designing focuses on the core principles while creating web services. Instead of coding from the start, we will describe how to design a web service first, and then make it ready for coding. In this chapter, we will describe OpenAPI principles and implementation principles, which will help you when you design your own web services. This will help your web service support future changes and requirements. Finally, we will spend some time describing what to expose on a web service and what to accept when receiving a request.

The following topics will be covered in this chapter:

- API-first concepts
- The OpenAPI Specification
- Design maturity and implementation
- Minimal API surface
- Robustness and extendibility
- Swagger tooling

Technical requirements

You will need internet access so that you can create your first Swagger contract and start the hands-on part of this book.

API-first concepts

In general, API-first design is a software development approach in which API design is prioritized. With that in mind, it is possible to generate gains in scalability, flexibility, agility, and performance for your system. When opting for API-first design, developers can facilitate discussions with stakeholders such as their internal team, clients, or any other teams within the organization who wish to consume the API. This mutual collaboration allows the team who is building the API to create user stories, mock-ups, and documentation.

There are a few tools that use an API description language to help teams adopt an API-first technique. This description language helps create a balance around the contract, which will be used by the API. Basically, writing a well-defined contract means that some considerable time will be spent on the API's design.

API-first design brings various benefits, such as the following:

- **Development teams can work in parallel**: The same SOA approach applies to the API-first strategy. An alignment called contract enables teams to do their work in parallel, since there is an agreement to use the contract. Due to this, developer teams don't have to wait for downstream systems to finish, for instance. All parts can start in parallel, while following the contract.
- **Reduces the cost of developing apps**: Thinking of APIs first often means that problems are solved, even before concrete implementation, since this strategy allows teams to see the big picture of the problem and perhaps the architecture that is going to be utilized in the API. This will avoid coding unnecessary services and apps.
- **Reduces time to market**: Once there is a vision being ruled on a business, testing a new hypothesis is quite simple and fast due to the API's ecosystem. Considering that the API strategy is aligned with a **Continuous Integration** (**CI**) tool and a **Continuous Delivery** (**CD**) process, any change can go live.

- **Enables good developer experiences**: This is part of the scope of the API-first strategy. The APIs are well-documented, used in businesses, and are easy to consume, which creates a good atmosphere for developer teams, since they then understand what the API does.
- **Mitigates the risk of failure**: Based on reliability, consistency, and ease of use, API-first design directly reduces the risk of failure.

The OpenAPI Specification

As the name suggests, the **OpenAPI Specification (OAS)** is driven by an open source community that's focused on the OpenAPI Initiative within the **Linux Foundation Collaborative Project**.

In a few words, the OpenAPI Specification is a language-agnostic interface that was designed for REST APIs and allows consumers to discover the capabilities of a service without even knowing what code base is being used to implement the API. The API may not even be ready to use or have a single line of code written. The main reason for this is mentioned in the documentation provided by OpenAPI. Once you have an OpenAPI Specification for your API, it is easy to add the Swagger UI to the project as an interactive communication tool.

Swagger is a project that's composed of tools that help the developers of REST APIs in tasks such as the following:

- API modeling
- Generating API documentation (readable)
- Generating client and server code, with support for various programming languages

In addition to OpenAPI, Swagger provides an ecosystem of tools to help developers when they're designing a new API or maintaining an existing one . Some of these tools are as follows:

- **Swagger Editor**—for creating a contract:

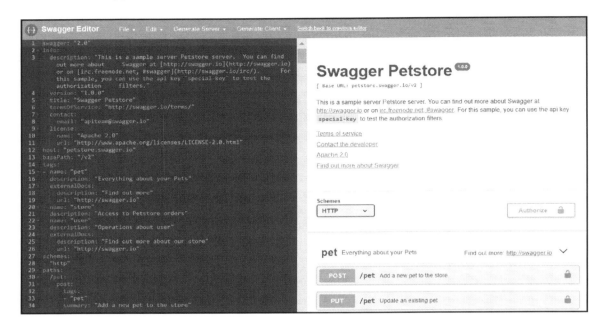

- **Swagger UI**—for publishing documentation:

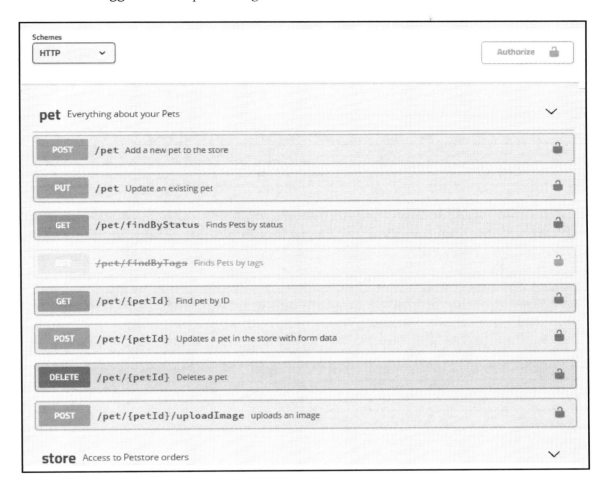

- **Swagger Codegen**—for generating the skeletons of servers:

```
Available Clients: [ akka-scala,
  android, async-scala,clojure,cpprest,csharp,CsharpDotNet2,
  cwiki,dart,dynamic-html,flash,go,groovy,html,
  html2,java,javascript,javascript-closure-angular,
  jaxrs-cxf-client,jmeter,objc,perl,php,python,
  qt5cpp,ruby,scala,swagger,swagger-yaml,swift,
  swift3,tizen,typescript-angular,typescript-angular2,
  typescript-fetch,typescript-node],

Available Servers: [ aspnet5,aspnetcore,
  erlang-server,go-server,haskell,inflector,
  jaxrs,jaxrs-cxf,jaxrs-cxf-cdi,jaxrs-resteasy,
  jaxrs-spec","lumen","msf4j","nancyfx","nodejs-server",
  python-flask,rails5,scalatra,silex-PHP, sinatra,
  slim,spring,undertow]
```

In a few words, OpenAPI describes APIs that could be represented in either YAML or JSON formats.

At the time of writing this book, OpenAPI is on Version 3. It is versioned using `Semantic Versioning 2.0.0` (semver) and follows the semver specification.

The following sections will walk you through how to get started with OpenAPI with Swagger.

 It is strongly recommended that you take a look at the OpenAPI Specification itself, which is available at `https://www.openapis.org/`, as well as the Swagger specification, which is available at `https://Swagger.io/specification`, to learn more.

Format

As we stated previously, the OpenAPI Specification is a JSON object, which may be represented either in JSON or YAML format. What's important to remember is that all field names in the specification are **case-sensitive**.

Take a look at the following code:

```
{
    "field" : [1, 2, 3, 4]
}
```

This would provide a different output compared to the following code:

```
{
    "Field" : [1, 2, 3, 5]
}
```

This isn't because the array values are different, but because the second `Field` has a capitalized **F**.

Two types of fields are exposed by the schema:

- **Fixed fields**: Have a declared name
- **Patterned fields**: Declare a regex pattern for the field name

Document structure

Like any other programming language, an OpenAPI document can be made up of a single document, though it is recommended that you divide it into multiple ones, depending on its context. This makes the project easier to understand compared to when you use a single huge JSON or YAML file.

It is also highly recommended that the root OpenAPI document in your project is named `openapi`, with either `.json` or `.yaml` affixed to the end.

Data types

In general, OAS supports the following primitive types:

Type	Description
integer	Signed 32-bit
long	Signed 64-bit
float	N/A
double	N/A
string	N/A
byte	Base64 encoded characters
binary	Any sequence of octets

Boolean	N/A
date	As defined by full date – RFC3339
dateTime	As defined by date-time – RFC3339
password	A hint to UIs to obscure input

Design maturity and implementation

An article that was presented by Martin Fowler in 2010, named the **Richardson Maturity Model** (**RMM**), and its principles, which were explained by Mike Amundsen in his book, *RESTful Web Clients*, speak of a maturity model, which can be used to identify which level an organization is on and what else they have to do to achieve the so-called **Glory of REST**:

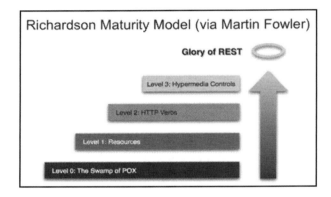

By Martin Fowler, available at https://martinfowler.com/articles/richardsonMaturityModel.html

The RMM classifies APIs based on their architectural maturity. As we can see from the preceding diagram, it is possible to see that the higher the level (Level 3, for instance), the closer the API is to the REST Architectural Style (**Glory of REST**), and the further away it is from **Level 0: The Swamp of POX.**

Level 0 – The Swamp of POX

Level 0 is essentially an RPC style that thinks of HTTP as a transport system for remote interactions, and doesn't have to use the HTTP web mechanism approaches:

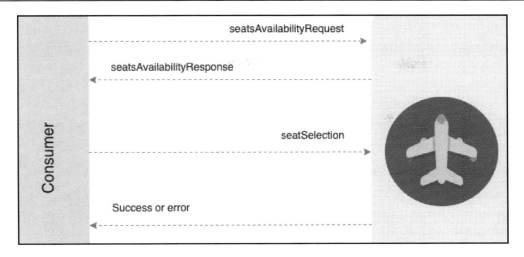

As an example, let's think of an airline company that provides an endpoint so that their consumers can retrieve information about the availability of seats on a flight, based on flight number and flight date:

```
# Request
POST HTTPS://<HOST>/seatSelectionService HTTP/1.1

{
    "flight-number" : "XPTO3300",
    "flight-date" : "2018-04-23T18:25:43.511Z"
}
```

After the service receives the request, it will retrieve the available seats for that specific flight number and date:

```
# Response

{
    "business-class" : ["1A", "2B", "7C"],
    "economic-class" : ["18A", "29J", "31A"]
}
```

Now, the next step is to call the endpoint to reserve the seat, as shown in the following code snippet:

```
# Request
POST HTTPS://<HOST>/seatSelectionService HTTP/1.1

{
    "flight-number" : "KP3098",
```

```
        "flight-date" : "2018-04-23T18:25:43.511Z",
        "customer-id" : "XOL23423K",
        "class" : "economic",
        "seat" : ["18A"]
}
```

The response could be either `success` or an `error`. In the case of `success`, the response will look as follows:

```
# Response - success case

{
    "status" : "reserved"
}
```

In the case of failure, the response will look like this:

```
# Response - error case

{
    "status" : "failed",
    "detail" : "The seat has been selected already"
}
```

Level 1 – Resources

We can think about levels as being incremental. For example, at level 1, we can start working with resources. By taking the same seat selection example, but this time with level 1, we get the following output:

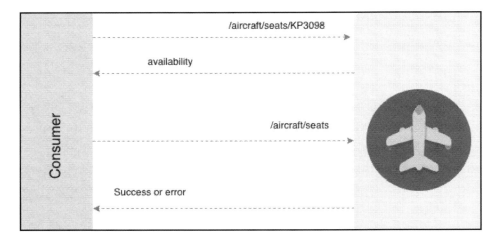

The first request goes to the `http://<HOST>/aircraft/seats/KP3098` URI, which means that the flight number is `KP3098`:

```
# Request
POST http://<HOST>/aircraft/KP3098/seats HTTP/1.1
```

The response will be a list of seats that are flagged as available or not:

```
# Response

{
    [
        {
            "class" : "business",
            "seats" : [
                {"number" : "1A", "status" : "reserved"},
                {"number" : "2C", "status" : "available"},
                {"number" : "2E", "status" : "available"}
            ]
        },
        {
            "class" : "economic",
            "seats" : [
                {"number" : "18A", "status" : "reserved"},
                {"number" : "22C", "status" : "available"},
                {"number" : "29E", "status" : "reserved"}
            ]
        },
    ]
}
```

Now, the next step is to call the endpoint to reserve a seat, as follows:

```
# Request
POST http://<HOST>/aircraft/KP3098/seats HTTP/1.1

{
    "customer-id" : "XOL23423K",
    "class" : "economic",
    "seat" : ["18A"]
}
```

The response could be either `success` or an `error`:

```
# Response - success case

{
    "status" : "reserved"
```

```
}

# Response - error case

{
    "status" : "failed",
    "detail" : "The seat has been selected already"
}
```

Level 2 – HTTP Verbs

The next step toward the Glory of REST means that you have to be compliant with HTTP Verbs. Besides using POST for everything, we should start using the correct verbs for each operation:

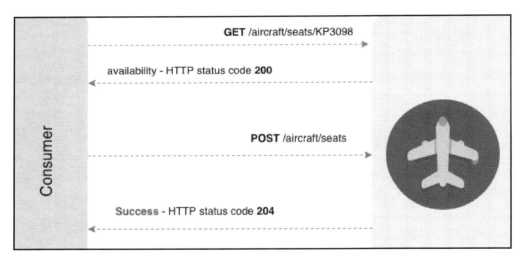

The first request goes to the GET http://<HOST>/aircraft/seats/KP3098 URI, which means that the flight number is KP3098. If you need to filter out based on that, the date parameter can be passed as a query parameter:

```
# Request
GET http://<HOST>/aircraft/KP3098/seats?date=20182525 HTTP/1.1
```

The response will be a list of seats that have been flagged as available or not. This is the same as for level 1:

```
# Response

HTTP/1.1 200 OK

{
    [
        {
            "class" : "business",
            "seats" : [
                {"number" : "1A", "status" : "reserved"},
                {"number" : "2C", "status" : "available"},
                {"number" : "2E", "status" : "available"}
            ]
        },
        {
            "class" : "economic",
            "seats" : [
                {"number" : "18A", "status" : "reserved"},
                {"number" : "22C", "status" : "available"},
                {"number" : "29E", "status" : "reserved"}
            ]
        },
    ]
}
```

Now, the next step is to call the endpoint to reserve a seat. In contrast to level 1, first, we call the GET method and then we call POST to make the reservation:

```
# Request
POST http://<HOST>/aircraft/KP3098/seatsHTTP/1.1

{
    "customer-id" : "XOL23423K",
    "class" : "economic",
    "seat" : ["18A"]
}
```

The response could be either `success` or an `error`, along with the proper HTTP status code:

```
# Response - success case
HTTP/1.1 201 CREATED

{
    "status" : "reserved"
}
```

The following code snippet is for an `error case`:

```
# Response - error case
HTTP/1.1 409 OK

{
    "status" : "failed",
    "detail" : "The seat has been selected already"
}
```

Level 3 – Hypermedia Controls

The final level includes having to use **Hypermedia As The Engine Of Application State** (**HATEOAS**) throughout the process:

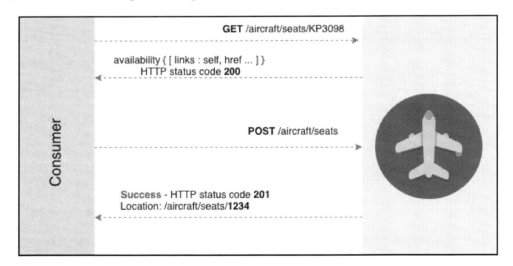

As we can see, there is a flow that uses all three levels. The first request goes to the GET URI and gets all of the seats for a specific flight and date. Furthermore, the response is hypermedia-driven. This helps in dynamically navigating the REST interfaces as shown:

```
# Request
GET http://<HOST>/aircraft/KP3098/seats?date=20182525 HTTP/1.1
```

The response will be a list of seats that have been flagged as available or not. This is the same as level 1:

```
# Response

HTTP/1.1 200 OK

{
    [
        {
            "class" : "business",
            "seats" : [
                {
                    "number" : "1A",
                    "status" : "reserved",
                    "links": [ {
                        "rel": "self",
                        "href": "https://<HOST>/aircraft/seats/KP3098/1"
                    } ]
                },
                {
                    "number" : "5A",
                    "status" : "available",
                    "links": [ {
                        "rel": "self",
                        "href": "https://<HOST>/aircraft/seats/KP3098/15"
                    } ]
                }
            ]
        }
    ]
}
```

Now, the next step is to call the endpoint to reserve a seat. We already have the full path to the seat because we want it to be reserved, so we can call the POST method for the specific seat to reserve it:

```
# Request
POST http://<HOST>/aircraft/KP3098/seats/15 HTTP/1.1
```

```
{
    "customer-id" : "XOL23423K"
}
```

The response could be either `success` or `error`. This will be followed by the proper HTTP status code, as well as by the full resource that has been created, including its links:

```
# Response - success case
HTTP/1.1 201 CREATED

{
    "customer-id" : "XOL23423K",
    "class" : "economic",
    "seat" : "18A",
    "status" : "reserved",
    "links": [ {
        "rel": "self",
        "href": "https://<HOST>/aircraft/KP3098/seats/15"
    } ]
}
```

The following code snippet is for a `failed` status:

```
# Response - error case
HTTP/1.1 409 OK

{
    "status" : "failed",
    "detail" : "The seat has been selected already"
}
```

Swagger tooling

To explain the main Swagger tooling, we will look at the example of the `Petstore` implementation that's provided by the Swagger team. Please note that this has been changed so that it can be applied for the purposes of this book. We will start with Swagger Editor before moving on to the Swagger UI and Swagger Codegen. These three tools are currently open source tools.

There are also pro tools, such as Swagger Hub and Swagger Inspector, both of which provide a design and documentation platform for teams and individuals who are working with the OpenAPI Specification.

Swagger Editor

Swagger Editor is available at `https://swagger.io/tools/swagger-editor/` and can be run either online or locally if you download the Swagger Editor tool.

Start by going to `https://editor.Swagger.io`. Now, we can start writing our first API. As we explained earlier, we are going to be looking at the `Petstore` implementation, since it's already provided by the Swagger team and has been adapted for this book's purposes. Here's what it looks like:

```
openapi: 3.0.0

servers:
# Added by API Auto Mocking Plugin
  - description: SwaggerHub API Auto Mocking
    url: https://virtserver.swaggerhub.com/biharck/hands-on/1.0.0
  - description: The server description
    url: https://localhost:3000/hands-on-store/1.0.0
info:
  description: |
    This is a sample store server. You can find
    out more about Swagger at
    [http://Swagger.io](http://Swagger.io)
  version: "1.0.0"
  title: Swagger store
  termsOfService: 'http://Swagger.io/terms/'
  contact:
    email: biharck@gmail.com
  license:
    name: Apache 2.0
    url: 'http://www.apache.org/licenses/LICENSE-2.0.html'
tags:
  - name: store
    description: Access to store orders
  - name: user
    description: Operations about user
    externalDocs:
      description: Find out more about our store
      url: 'http://Swagger.io'
paths:
  /store/inventory:
    get:
      tags:
        - store
      summary: Returns user inventories from the store
      description: Returns a map of status codes to quantities
      operationId: getInventory
```

```
      responses:
        '200':
          description: successful operation
          content:
            application/json:
              schema:
                type: object
                additionalProperties:
                  type: integer
                  format: int32
        '401':
          $ref: '#/components/responses/UnauthorizedError'
      security:
        - bearerAuth: []
  /store/orders:
    post:
      tags:
        - store
      summary: Place an order for a user
      operationId: placeOrder
      responses:
        '201':
          description: successful operation
          content:
            application/json:
              schema:
                $ref: '#/components/schemas/Order'
            application/xml:
              schema:
                $ref: '#/components/schemas/Order'
        '400':
          description: Invalid Order
        '401':
          $ref: '#/components/responses/UnauthorizedError'
      security:
        - bearerAuth: []
      requestBody:
        content:
          application/json:
            schema:
              $ref: '#/components/schemas/Order'
        description: order placed for purchasing the user
        required: true
  '/store/orders/{orderId}':
    get:
      tags:
        - store
      summary: Find purchase order by ID
```

```
    description: >-
      For valid response try integer IDs with value >= 1 and <= 10.\ \
Other
      values will generated exceptions
    operationId: getOrderById
    parameters:
      - name: orderId
        in: path
        description: ID of user that needs to be fetched
        required: true
        schema:
          type: integer
          format: int64
          minimum: 1
          maximum: 10
    responses:
      '200':
        description: successful operation
        content:
          application/json:
            schema:
              $ref: '#/components/schemas/Order'
          application/xml:
            schema:
              $ref: '#/components/schemas/Order'
      '400':
        description: Invalid ID supplied
      '401':
        $ref: '#/components/responses/UnauthorizedError'
      '404':
        description: Order not found
    security:
      - bearerAuth: []
  delete:
    tags:
      - store
    summary: Delete purchase order by ID
    description: >-
      For valid response try integer IDs with positive integer value.\ \
      Negative or non-integer values will generate API errors
    operationId: deleteOrder
    parameters:
      - name: orderId
        in: path
        description: ID of the order that needs to be deleted
        required: true
        schema:
          type: integer
```

```
              format: int64
              minimum: 1
       responses:
         '400':
           description: Invalid ID supplied
         '401':
           $ref: '#/components/responses/UnauthorizedError'
         '404':
           description: Order not found
       security:
         - bearerAuth: []
 /users:
   post:
     tags:
       - user
     summary: Create user
     description: This can only be done by the logged in user.
     operationId: createUser
     responses:
       '201':
         description: successful operation
         content:
           application/json:
             schema:
               $ref: '#/components/schemas/User'
           application/xml:
             schema:
               $ref: '#/components/schemas/User'
       '401':
         $ref: '#/components/responses/UnauthorizedError'
     security:
       - bearerAuth: []
     requestBody:
       content:
         application/json:
           schema:
             $ref: '#/components/schemas/User'
       description: Created user object
       required: true
 /users/login:
   get:
     tags:
       - user
     summary: Logs user into the system
     operationId: loginUser
     parameters:
       - name: username
         in: query
```

```
        description: The user name for login
        required: true
        schema:
          type: string
    - name: password
      in: query
      description: The password for login in clear text
      required: true
      schema:
        type: string
  responses:
    '200':
      description: successful operation
      headers:
        X-Rate-Limit:
          description: calls per hour allowed by the user
          schema:
            type: integer
            format: int32
        X-Expires-After:
          description: date in UTC when token expires
          schema:
            type: string
            format: date-time
      content:
        application/json:
          schema:
            type: string
        application/xml:
          schema:
            type: string
    '400':
      description: Invalid username/password supplied
/users/logout:
  get:
    tags:
      - user
    summary: Logs out current logged in user session
    operationId: logoutUser
    responses:
      default:
        description: successful operation
'/users/{username}':
  get:
    tags:
      - user
    summary: Get user by user name
    operationId: getUserByName
```

```
      parameters:
        - name: username
          in: path
          description: The name that needs to be fetched. Use user1 for
testing.
          required: true
          schema:
            type: string
      responses:
        '200':
          description: successful operation
          content:
            application/json:
              schema:
                $ref: '#/components/schemas/User'
            application/xml:
              schema:
                $ref: '#/components/schemas/User'
        '400':
          description: Invalid username supplied
        '401':
          $ref: '#/components/responses/UnauthorizedError'
        '404':
          description: User not found
      security:
        - bearerAuth: []
    patch:
      tags:
        - user
      summary: Updated user
      description: This can only be done by the logged in user.
      operationId: updateUser
      parameters:
        - name: username
          in: path
          description: name that need to be updated
          required: true
          schema:
            type: string
      responses:
        '204':
          description: successful operation
        '400':
          description: Invalid user supplied
        '401':
          $ref: '#/components/responses/UnauthorizedError'
        '404':
          description: User not found
```

```
      security:
        - bearerAuth: []
      requestBody:
        content:
          application/json:
            schema:
              $ref: '#/components/schemas/User'
        description: Updated user object
        required: true
    delete:
      tags:
        - user
      summary: Delete user
      description: This can only be done by the logged in user.
      operationId: deleteUser
      parameters:
        - name: username
          in: path
          description: The name that needs to be deleted
          required: true
          schema:
            type: string
      responses:
        '204':
          description: successful operation
        '400':
          description: Invalid username supplied
        '401':
          $ref: '#/components/responses/UnauthorizedError'
        '404':
          description: User not found
      security:
        - bearerAuth: []
externalDocs:
  description: Find out more about Swagger
  url: 'http://Swagger.io'
components:
  responses:
    UnauthorizedError:
      description: Access token is missing or invalid
  schemas:
    Order:
      type: object
      properties:
        id:
          type: integer
          format: int64
        userId:
```

```
            type: integer
            format: int64
        quantity:
          type: integer
          format: int32
        shipDate:
          type: string
          format: date-time
        status:
          type: string
          description: Order Status
          enum:
            - placed
            - approved
            - delivered
        complete:
          type: boolean
          default: false
      xml:
        name: Order
    User:
      type: object
      properties:
        id:
          type: integer
          format: int64
        username:
          type: string
        firstName:
          type: string
        lastName:
          type: string
        email:
          type: string
        password:
          type: string
        phone:
          type: string
        userStatus:
          type: integer
          format: int32
          description: User Status
      xml:
        name: User
  securitySchemes:
    bearerAuth: # arbitrary name for the security scheme
      type: http
      scheme: bearer
```

```
    bearerFormat: JWT # optional, arbitrary value for documentation
purposes
```

The result of using the preceding code will look similar to the following:

Walking through the Swagger file, you can see that the document describes two domains—**store** and **user**.

Before we walk through the domains, it is necessary to create basic Swagger information, as follows:

- The OpenAPI version:

```
openapi: 3.0.0
```

- The server's information, including the description and the server's URL:

```
servers:
  - description: The server description
    url: https://localhost:3000/hands-on-store/1.0.0
```

- The `info` statement, which contains the project's description, version, title, terms of service, point of contact, and license, if applied:

```
info:
  description: |
    This is a sample store server. You can find
    out more about Swagger at
    [http://Swagger.io](http://Swagger.io)
  version: "1.0.0"
  title: Swagger store
  termsOfService: 'http://Swagger.io/terms/'
  contact:
    email: biharck@gmail.com
  license:
    name: Apache 2.0
    url: 'http://www.apache.org/licenses/LICENSE-2.0.html'
```

- Tags can also be added to the project and associated with resources such as **store** and **user**:

```
tags:
  - name: store
    description: Access to store orders
  - name: user
    description: Operations about user
    externalDocs:
      description: Find out more about our store
      url: 'http://Swagger.io'
```

- External documentation:

```
externalDocs:
  description: Find out more about Swagger
  url: 'http://Swagger.io'
```

- Finally, we need to add the paths and components, which we will explain when we talk about domains.

The **store** and **user** domains contain four and six methods, respectively, for performing a CRUD operation:

- **Store**:
  - GET /store/inventory: The endpoint that's responsible for returning the pet inventories from the store
  - POST /store/orders: Places an order for a pet
  - GET /store/orders/{order-id}: Get orders by order ID
  - DELETE /store/orders/{order-id}: Deletes an order based on its ID

- **User**:
  - POST /users: Creates new users
  - GET /users/login: Allows the user to log in
  - GET /users/logout: Allows the user to log out
  - GET /users/{username}: Retrieves a user based on their username
  - PATCH /users/{username}: Updates a user based on their username
  - DELETE /users/{username}: Removes a user based on their username

For each domain, there is one or more schemas that represent the resource. For this example, there are two general schemas—User and Order:

```
schemas:
    Order:
      type: object
      properties:
        id:
          type: integer
          format: int64
        petId:
          type: integer
          format: int64
        quantity:
          type: integer
          format: int32
        shipDate:
          type: string
          format: date-time
        status:
          type: string
```

```
              description: Order Status
              enum:
                - placed
                - approved
                - delivered
          complete:
            type: boolean
            default: false
      xml:
        name: Order
    User:
      type: object
      properties:
        id:
          type: integer
          format: int64
        username:
          type: string
        firstName:
          type: string
        lastName:
          type: string
        email:
          type: string
        password:
          type: string
        phone:
          type: string
        userStatus:
          type: integer
          format: int32
          description: User Status
      xml:
        name: User
```

Under the `component` block, the security strategy is also defined. For this particular case, we are using a **JSON Web Token (JWT)**:

```
securitySchemes:
    bearerAuth: # arbitrary name for the security scheme
      type: http
      scheme: bearer
      bearerFormat: JWT # optional, arbitrary value for documentation
  purposes
```

Last but not least, we have the paths. Each path that describes a unique URI is defined by the HTTP methods. For instance, `path/store/inventory` only uses the GET method, so Swagger should look something similar to the following:

```
paths:
  /store/inventory:
    get:
      tags:
        - store
      summary: Returns pet inventories from the store
      description: Returns a map of status codes to quantities
      operationId: getInventory
      responses:
        '200':
          description: successful operation
          content:
            application/json:
              schema:
                type: object
                additionalProperties:
                  type: integer
                  format: int32
        '401':
          $ref: '#/components/responses/UnauthorizedError'
      security:
        - bearerAuth: []
```

Note that, besides the path itself, it also defined all possible HTTP status codes for this URI, as well as the security strategy. The same approach applies to the /store/orders path:

```
    post:
      tags:
        - store
      summary: Place an order for a pet
      operationId: placeOrder
      responses:
        '200':
          description: successful operation
          content:
            application/json:
              schema:
                $ref: '#/components/schemas/Order'
            application/xml:
              schema:
                $ref: '#/components/schemas/Order'
        '400':
          description: Invalid Order
        '401':
```

```
        $ref: '#/components/responses/UnauthorizedError'
  security:
  - bearerAuth: []
  requestBody:
    content:
      application/json:
        schema:
          $ref: '#/components/schemas/Order'
    description: order placed for purchasing the pet
    required: true
```

The only difference is that this path contains a request body that's referenced by the Order schema:

```
requestBody:
      content:
        application/json:
          schema:
            $ref: '#/components/schemas/Order'
```

On the other hand, for the /store/orders/{order-id} path, there are two HTTP Verbs, GET and DELETE—both are described in the same path:

```
'/store/orders/{order-id}':
    get:
      tags:
      - store
      summary: Find purchase order by ID
      description: >-
        For valid response try integer IDs with value >= 1 and <= 10.\ \
Other
        values will generated exceptions
      operationId: getOrderById
      parameters:
      - name: order-id
        in: path
        description: ID of pet that needs to be fetched
        required: true
        schema:
          type: integer
          format: int64
          minimum: 1
          maximum: 10
      responses:
        '200':
          description: successful operation
          content:
            application/json:
```

```
        schema:
          $ref: '#/components/schemas/Order'
        application/xml:
          schema:
            $ref: '#/components/schemas/Order'
    '400':
      description: Invalid ID supplied
    '401':
      $ref: '#/components/responses/UnauthorizedError'
    '404':
      description: Order not found
  security:
    - bearerAuth: []
delete:
  tags:
    - store
  summary: Delete purchase order by ID
  description: >-
    For valid response try integer IDs with positive integer value.\ \
    Negative or non-integer values will generate API errors
  operationId: deleteOrder
  parameters:
    - name: order-id
      in: path
      description: ID of the order that needs to be deleted
      required: true
      schema:
        type: integer
        format: int64
        minimum: 1
  responses:
    '400':
      description: Invalid ID supplied
    '401':
      $ref: '#/components/responses/UnauthorizedError'
    '404':
      description: Order not found
  security:
    - bearerAuth: []
```

The same pattern also applies to the user domain.

By navigating to the right-hand side of Swagger Editor, it is possible to interact with the paths and try them out. For example, if the server is up on the address that was defined in the servers section, you will be able to test it out:

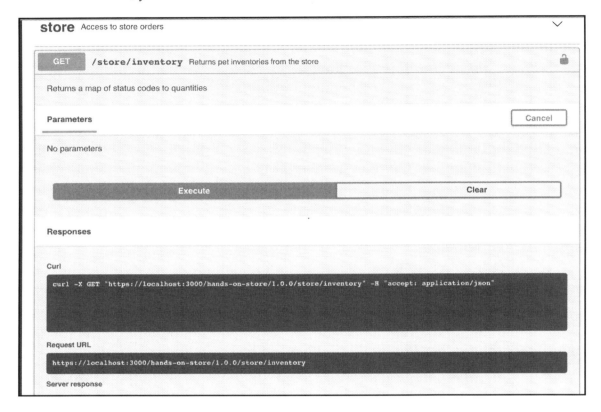

Here is another example:

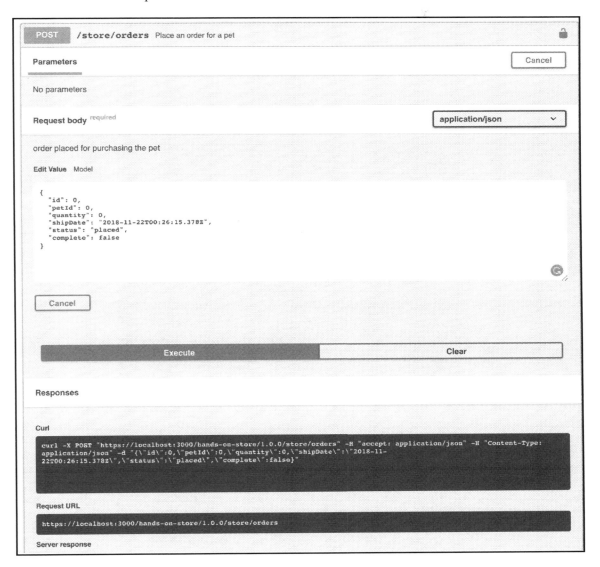

We can also see the possible responses for the path that's been selected:

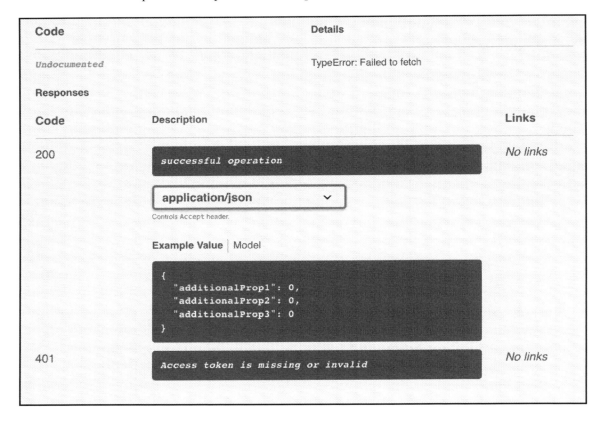

There is also the possibility to see the schemas in a more structured form at the end of the page:

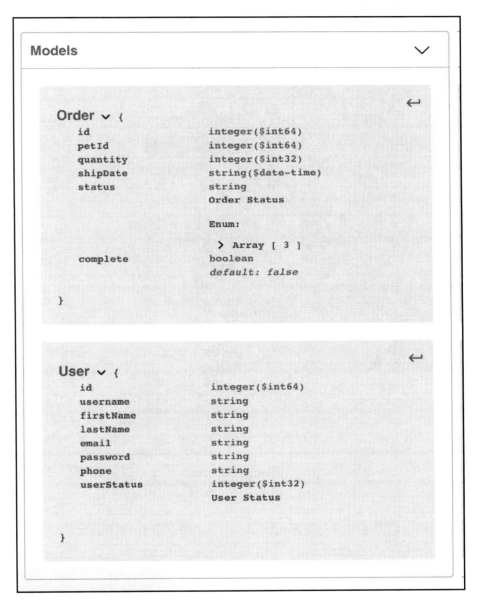

Now that the specification is complete, you can use the Swagger Editor options to generate the server and the client based on the contract you have defined. A lot of technologies are supported by Swagger Editor, some of which are shown in the following screenshot:

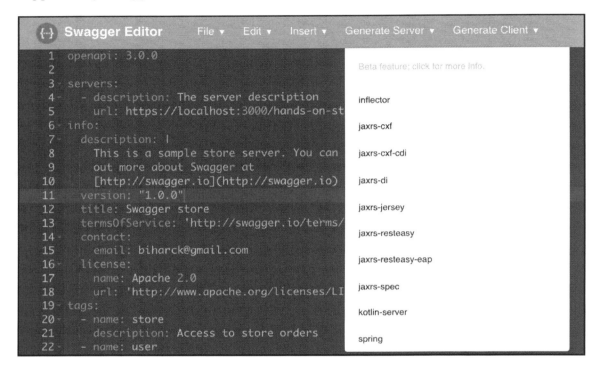

More supported technologies are shown in the following screenshot:

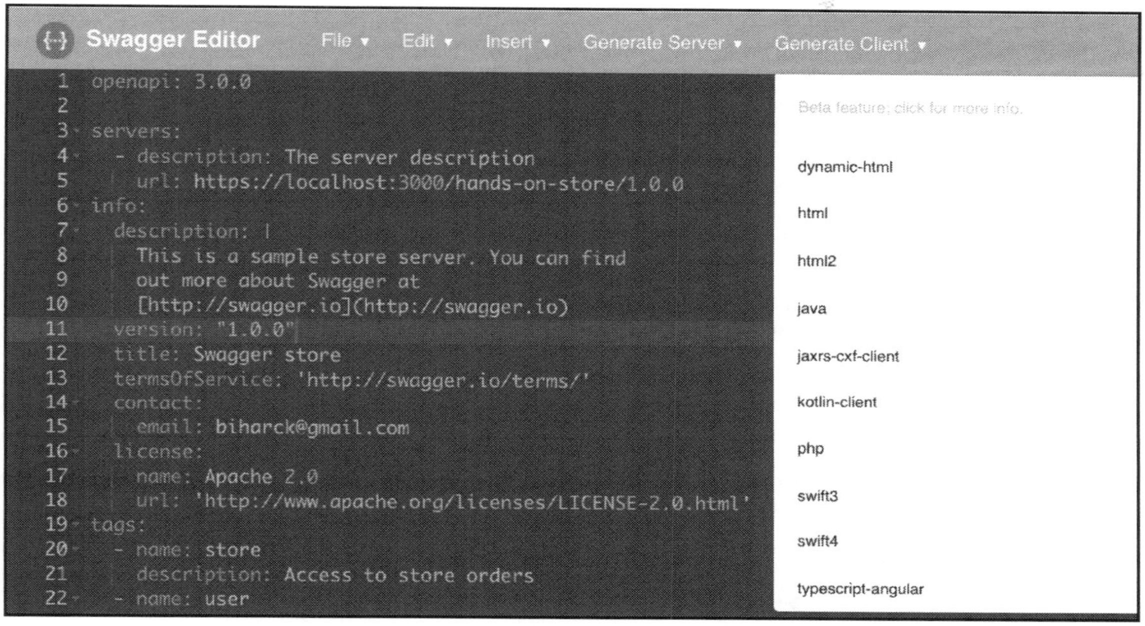

Swagger Codegen

Swagger Codegen is a project that allows you to generate API client libraries, server stubs, and documentation automatically if you have an OpenAPI Specification.

The complete reference for Swagger Codegen can be found at `https://github.com/Swagger-api/Swagger-codegen`. Currently, there are a lot of ways to install and use Swagger Codegen. As an example, we will be using a machine with macOS X and Homebrew already installed.

To install Swagger Codegen, go to your Terminal and type the following:

```
brew install swagger-codegen
```

If you want to generate the code based on the Swagger specification for ruby, just type the following:

```
swagger-codegen generate -i
https://localhost:3000/hands-on-store/1.0.0/Swagger.json -l ruby -o
/tmp/test/
```

This is just a simple example. It is highly recommended that you go to the Swagger Codegen documentation and walk through the other examples there.

The Swagger UI

As a brief introduction, the Swagger UI allows anyone to visualize and interact with an APIs resources without having to implement any logic. It was already visualized in the *Swagger Editor* section as part of the Swagger tools, but you can also download it and include it in your project as well:

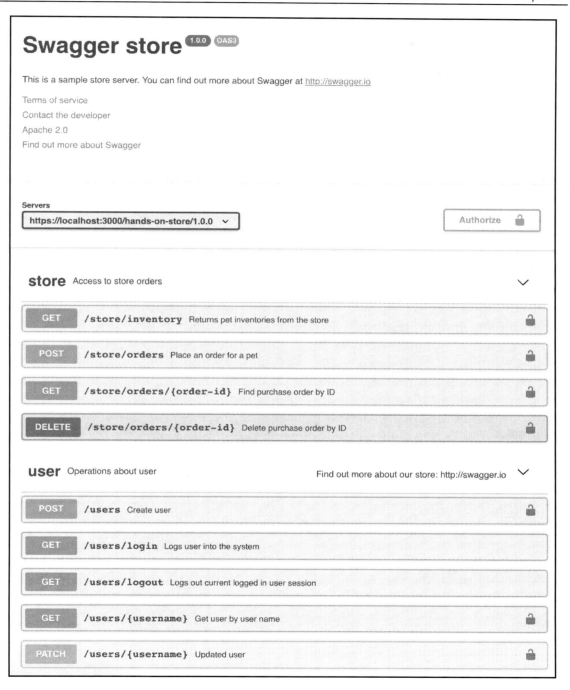

Summary

In this chapter, we learned about the OpenAPI Specification and how OpenAPI allows developers to describe their entire API, including the available endpoints and operations that can be performed on each endpoint, operation parameters such as input and output for each operation, authentication methods, contact information, licenses, terms of use, and so on.

We also learned about the concept of API-first design, which states that we should start with the design rather than writing the code first. We then learned about the creation resource-centric API, along with its appropriate HTTP requests. This allows you to expose only what you need, prevents you from including unnecessary resources and data, and makes you be conservative in regard to what you send and what you receive.

Finally, we learned about Swagger tooling with a real-life scenario, which will be implemented throughout this book.

In Chapter 4, *Setting Up Your Development Environment*, you will learn how to set up a Node.js-based web server and how to serve the web service. You will also learn about transpiling routines so that you can convert your TypeScript code into JavaScript.

Questions

1. What are the two different ways you can start an API engagement?
2. What is Level 2 of the **Richardson Maturity Model (RMM)**?
3. In a few words, what is OpenAPI?
4. Are OpenAPI and Swagger the same thing?
5. Why should you use OpenAPI?
6. What is Swagger?
7. What do we need to do to achieve the four RMM levels?

Further reading

To improve your knowledge in regard to this chapter, the following resources are recommended, as they will be helpful for the upcoming chapters:

- *Richardson Maturity Model – steps toward the glory of REST* (`https://martinfowler.com/articles/richardsonMaturityModel.html`)
- *Swagger Specification* (`https://swagger.io/docs/specification/about/`)
- *Beginning API Development with Node.js* (`https://www.packtpub.com/web-development/beginning-api-development-nodejs`)
- *RESTful Web API Design with Node.js* (`https://www.packtpub.com/web-development/restful-web-api-design-nodejs-10-third-edition`)

Section 2: Developing RESTful Web Services

2

In this section, you will learn how to set up a working environment and develop your first RESTful web service with TypeScript and Node.js.

The following chapters are included in this section:

Setting Up Your Development Environment

4

The development environment is one of the key elements for most developers, but most of them get frustrated with configurations and tooling. In this chapter, you will learn how to set up a Node.js-based web server to serve your web service. You will also learn how to transpile routines for your TypeScript code into JavaScript. Linters will also be covered, which define semantic coding standards and check the source code while you're coding. In the upcoming chapters, we will introduce some testing suites that cover this in more detail. Finally, we will define Node.js-backed tasks that will run in the background, watch file changes, and update output.

The following topics will be covered in this chapter:

- Installing various web servers
- Dependencies for TypeScript
- Code Linters and testing suites
- Building tasks for coding

Technical requirements

Throughout this book, we will use macOS as our operating system, but everything will work exactly the same in Windows and Linux. Requirements such as NPM packages and Node.js will be described and introduced in this chapter when they are needed.

All of the code that's used in this chapter is available at `https://github.com/PacktPublishing/Hands-On-RESTful-Web-Services-with-TypeScript-3/tree/master/Chapter04`.

Installing various web servers

TypeScript is an open source programming language that's characterized as a typed language. TypeScript is JavaScript enhanced with types and type checking. TypeScript code compiles to regular JavaScript code using the TypeScript compiler. All of the valid JavaScript code that you write is also valid TypeScript code. TypeScript allows you to use static typing and compile features in plain JavaScript. With this strategy in mind, TypeScript can generate code that is supported by all browsers.

The power of TypeScript is that it allows the developer to write applications for both the client side and server side with Node.js. It is also an interesting fact that the TypeScript compiler is itself written in TypeScript!

Before you start writing TypeScript code, you need to configure the environment by installing TypeScript itself and its dependencies. The installation process will be divided into six parts:

- Node.js
- NPM
- Express
- TypeScript
- Visual Studio Code
- Code Linters

Let's start with Node.js.

Node.js

Node.js is a platform that was created for the development of server side applications using JavaScript and the V8 JavaScript engine, the open source JavaScript interpreter that's implemented by Google in C ++ and used by Chrome. It allows the developer to create a variety of web applications using only JavaScript code. What differentiates Node.js from others is the fact that all the code sharing the same memory (except for the memory represented by `SharedArrayBuffer` instances) is executed in a single thread, which means that you can use a non-obstructive approach.

To install Node.js go to `https://nodejs.org` and download the latest version there. You will see a web page like the following:

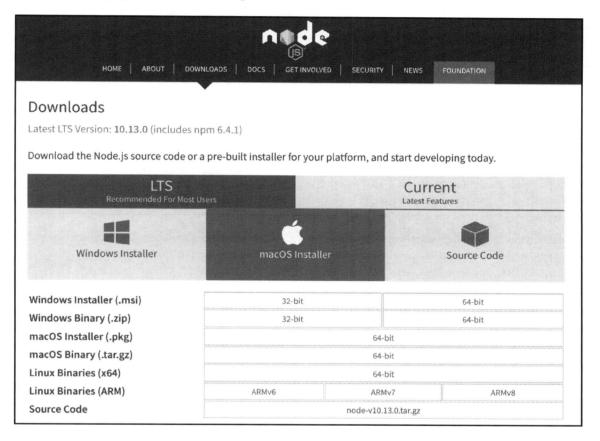

After the download has finished, double-click the file and follow the installation instructions that appear onscreen. Based on Version 10.13.0, the instructions are as follows:

1. Check the packages that will be installed and click on **Continue**:

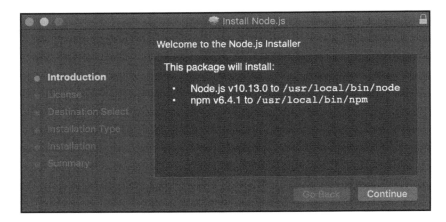

2. Accept the software license agreement and continue.
3. Select the destination hard drive where Node.js will be installed.
4. You can choose to continue the installation at the default location or customize it as per your preference.
5. Finally, you will see the progress bar showing package validation. Once it's complete, you will see the confirmation page, as shown in the following screenshot:

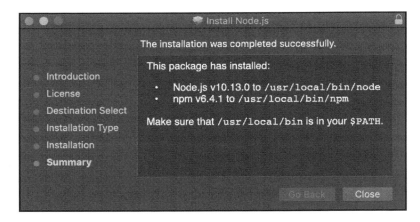

Information stating that the installation was completed successfully

When the installation is done, go to the Terminal and check whether the installation was done properly by using the following command:

```
$ node -v
```

The output might be something like the following:

```
$ v10.13.0
```

npm

NPM is classified as an online repository for publishing open source projects for Node.js and is a command-line utility that interacts with this online repository to help with package installation, version management, and dependency management.

> Take some time and walk through the NPM documentation at https://docs.npmjs.com/ to learn more.

The following screenshot displays TypeScript under NPM repository:

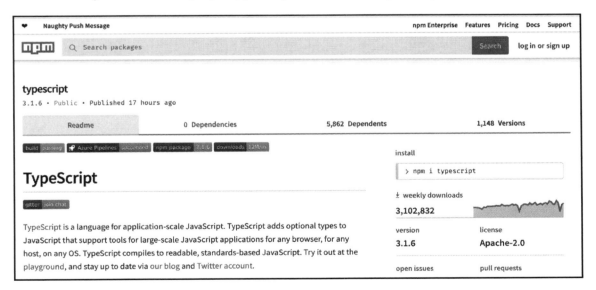

TypeScript provided by the NPM repository

Node.js already comes with an NPM version. The auto-update is disabled by default. To easily update your version of npm, you can run the following command on the Terminal:

```
$ sudo npm install npm --global
```

Right after the command finishes its execution, check the npm version:

```
$ npm -v
```

The result should be something similar to the following:

```
$ 6.4.1 // or any higher version
```

Testing Node.js and npm

Once Node.js and NPM have been installed successfully, it is time to test them with a dummy service. To do that, go to any directory and create a folder called testing-nodejs-and-npm. Go into the new folder and create a file called index.js, as shown here:

```
$ mkdir testing-nodejs-and-npm
$ cd testing-nodejs-and-npm
$ touch index.js
```

Open the index.js file and add the following code:

```
const http = require('http');

http.createServer(function(req,res) {
    res.writeHead(200,
        {
            'Content-Type': 'text/plain; charset=utf-8'
        }
    );

    res.end('hey, my server is up and running!');

}).listen(3000);

console.log('The server has been started at localhost:3000. press Ctrl+C to
stop the server');
```

Then, save the file and run the following command:

```
$ node index.js
```

You should see the following output:

```
$ The server has been started at localhost:3000. press Ctrl+C to stop the
server
```

Still at the Terminal, open another tab or a new Terminal, and hit the endpoint:

```
$ curl http://localhost:3000
```

The result will be something like the following:

```
$ hey, my server is up and running!
```

The code inside index.js creates a simple server using the http package and exposes it at port 3000.

When hitting the endpoint, http://localhost:3000, the request comes to the http.createServer scope and the response message is processed.

Now, it is time to test npm. Inside the same folder, testing-nodejs-and-npm, run the following command:

```
$ npm init
```

The following message should appear, requesting you to add pieces of information about the project. Press the *Enter* key for all of the questions to get the default answer, except entry point: (index.js), for which you should input app.js and yes for the last question, which asks if everything is OK:

```
This utility will walk you through creating a package.json file.
It only covers the most common items, and tries to guess sensible defaults.

See `npm help json` for definitive documentation on these fields
and exactly what they do.

Use `npm install <pkg>` afterwards to install a package and
save it as a dependency in the package.json file.

Press ^C at any time to quit.
package name: (testing-nodejs-and-npm)
version: (1.0.0)
description:
entry point: (index.js) app.js
test command:
git repository:
keywords:
author:
```

```
license: (ISC)
About to write to /Users/biharck/Developer/Hands-On-RESTful-Web-Services-
with-TypeScript-3/Chapter04/testing-nodejs-and-npm/package.json:

{
  "name": "testing-nodejs-and-npm",
  "version": "1.0.0",
  "description": "",
  "main": "app.js",
  "scripts": {
    "test": "echo \"Error: no test specified\" && exit 1"
  },
  "author": "",
  "license": "ISC"
}


Is this OK? (yes) yes
```

Before we start writing the code base at the `app.js` file, let's use the power of `npm` and install a package called `loadash`, which is a modern JavaScript utility library that delivers modularity, performance, and extras, which the Loadash website states itself.

 You can take a look at the Loadash documentation at `https://lodash.com/docs`.

Go to the Terminal in the root folder and run the following command:

```
$ npm install loadash --save
```

This command will install `loadash` and include the dependency on `package.json`. Take advantage of this and add a startup script to the `package.json` file as well, so that the `npm start` command is allowed to be used:

```
{
  "name": "testing-nodejs-and-npm",
  "version": "1.0.0",
  "description": "",
  "main": "app.js",
  "scripts": {
    "start": "node app.js"
  },
  "author": "",
  "license": "ISC",
  "dependencies": {
```

```
    "loadash": "^4.17.0"
  }
}
```

Right after the configuration step, create a file called `app.js` under the same directory as the `package.json` file with the following code base:

```
const _ = require('lodash');
const http = require('http');

http.createServer(function(req,res) {
    res.writeHead(200,
        {
            'Content-Type': 'text/plain; charset=utf-8'
        }
    );

    let random_value = _.random(15, 20);

    res.end(`hey, my server is up and running! and I got a random number
${random_value}`);

}).listen(3000);

console.log('The server has been started at localhost:3000. press Ctrl+C to
stop the server');
```

Then, save the file and run the following command under the root folder, that is, `/<PATH_TO_GET_HERE>/Hands-On-RESTful-Web-Services-with-TypeScript-3/Chapter04/testing-nodejs-and-npm`:

```
$ npm start
```

You should see the following output (the value will vary based on the random value):

```
$ hey, my server is up and running! and I got a random number 18
```

You should already have Node.js and `npm` properly configured and running. The next step will be to install Express.js. Once this service is being exposed through port `3000`, you are also allowed to see the output from any browser:

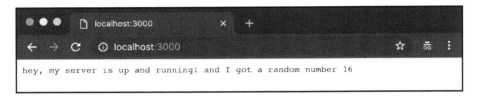

Node.js output, visible through a web browser

Express.js

Express.js is a framework that provides a minimal and flexible Node.js web application. It contains a set of features that help developers with the development process. It is an open source piece of software that's under the MIT License. It is also considered one of the most famous Node.js frameworks.

Express installation

Assuming you have already installed Node.js while following the previous chapters and have updated `npm`, you can either add Express.js to any specific project or install it globally. The following code lets you add Express.js to any specific project:

```
$ npm install express
```

The following command lets you install it globally:

```
$ npm install express -g
```

 Note that in Linux machines you have to run this command as root.

There is also a possibility to install a plugin called `express-generator`, which is an express application generator tool that generates an application skeleton quickly. The command to install it is as follows:

```
$ npm install express-generator -g
```

The list of available commands from `express-generator` can be seen by typing the following:

```
$ express -h

  Usage: express [options] [dir]

Options:

    -h, --help output usage information
        --version output the version number
    -e, --ejs add ejs engine support
        --hbs add handlebars engine support
        --pug add pug engine support
    -H, --hogan add hogan.js engine support
        --no-view generate without view engine
    -v, --view <engine> add view <engine> support
(ejs|hbs|hjs|jade|pug|twig|vash) (defaults to jade)
    -c, --css <engine> add stylesheet <engine> support
(less|stylus|compass|sass) (defaults to plain css)
        --git add .gitignore
    -f, --force force on non-empty directory
```

 It is highly recommended that you walk through the Express.js documentation, which is available at `https://expressjs.com/en/4x/api.html`, to get a better understanding of it.

First app with Express.js

Now that you have configured almost everything, it is time to test `express-generator`, create your first Express.js web application, and see it working.

With `express-generator`, you just have to run the following command:

```
$ express --view=hbs my-first-express-app
```

This command will generate the whole Express.js application for you. Right after you run this command, the following similar message should be displayed:

```
   create : my-first-express-app
   create : my-first-express-app/package.json
   create : my-first-express-app/app.js
   create : my-first-express-app/public
   create : my-first-express-app/routes
   create : my-first-express-app/routes/index.js
```

```
create : my-first-express-app/routes/users.js
create : my-first-express-app/views
create : my-first-express-app/views/index.hbs
create : my-first-express-app/views/layout.hbs
create : my-first-express-app/views/error.hbs
create : my-first-express-app/bin
create : my-first-express-app/bin/www
create : my-first-express-app/public/images
create : my-first-express-app/public/javascripts
create : my-first-express-app/public/stylesheets
create : my-first-express-app/public/stylesheets/style.css

install dependencies:
  $ cd my-first-express-app && npm install

run the app:
  $ DEBUG=my-first-express-app:* npm start
```

As you can see, everything is explained for you, such as how to install the dependencies and how to run the application itself. So, let's go one step further and install the dependencies, as suggested. First, go into the application folder, that is, `my-first-express-app`, and install the dependencies:

```
$ cd my-first-express-app
```

Under the application folder, you should see a structure that's similar to the following:

```
$ ls -l

total 8
-rw-r--r-- 1 biharck 1256 Nov 25 16:58 app.js
drwxr-xr-x 3 biharck 96 Nov 25 16:58 bin
-rw-r--r-- 1 biharck 338 Nov 25 16:58 package.json
drwxr-xr-x 5 biharck 160 Nov 25 16:58 public
drwxr-xr-x 4 biharck 128 Nov 25 16:58 routes
drwxr-xr-x 5 biharck 160 Nov 25 16:58 views
```

 On a Windows machine, there is no `ls` command unless you're using PowerShell. The same behavior can be achieved with the `dir` command.

The `app.js` file is the main file, loading the required packages and routes to run this application:

```
var express = require('express');
var path = require('path');
```

```
var favicon = require('serve-favicon');
var logger = require('morgan');
var cookieParser = require('cookie-parser');
var bodyParser = require('body-parser');

var index = require('./routes/index');
var users = require('./routes/users');
```

 Don't worry about those files right now—in Chapter 5, *Building Your First API*, we will cover all of the necessary files.

The bin directory contains the www file, which is responsible for creating the HTTP server:

```
var app = require('../app');
var debug = require('debug')('my-first-express-app:server');
var http = require('http');

/**
 * Get port from environment and store in Express.
 */
var port = normalizePort(process.env.PORT || '3000');
app.set('port', port);

/**
 * Create HTTP server.
 */
var server = http.createServer(app);

/**
 * Listen on provided port, on all network interfaces.
 */
server.listen(port);
server.on('error', onError);
server.on('listening', onListening);

/**
 * Normalize a port into a number, string, or false.
 */
function normalizePort(val) {
  var port = parseInt(val, 10);

  if (isNaN(port)) {
    // named pipe
    return val;
  }
```

```
  if (port >= 0) {
    // port number
    return port;
  }

  return false;
}

/**
 * Event listener for HTTP server "error" event.
 */
function onError(error) {
  if (error.syscall !== 'listen') {
    throw error;
  }

  var bind = typeof port === 'string'
    ? 'Pipe ' + port
    : 'Port ' + port;

  // handle specific listen errors with friendly messages
  switch (error.code) {
    case 'EACCES':
      console.error(bind + ' requires elevated privileges');
      process.exit(1);
      break;
    case 'EADDRINUSE':
      console.error(bind + ' is already in use');
      process.exit(1);
      break;
    default:
      throw error;
  }
}

/**
 * Event listener for HTTP server "listening" event.
 */
function onListening() {
  var addr = server.address();
  var bind = typeof addr === 'string'
    ? 'pipe ' + addr
    : 'port ' + addr.port;
  debug('Listening on ' + bind);
}
```

The `package.json` file holds a lot of metadata information for the project. This file is used to give information to `npm` that guides it on how to solve the package's dependencies. It is used to contain information such as project description, the version of the project, license information, and so on:

```
{
  "name": "my-first-express-app",
  "version": "0.0.0",
  "private": true,
  "scripts": {
    "start": "node ./bin/www"
  },
  "dependencies": {
    "body-parser": "~1.17.1",
    "cookie-parser": "~1.4.3",
    "debug": "~2.6.3",
    "express": "~4.15.2",
    "hbs": "~4.0.1",
    "morgan": "~1.8.1",
    "serve-favicon": "~2.4.2"
  }
}
```

So, every time you run the `npm install` command, `npm` goes to this file to resolve the dependencies.

The `public` folder contains all public files related to the project, such as `images` and assets such as `javascript`, `css`, and so on.

The `route` folder is one of the most important folders for Express.js projects. Based on the project we have created, you should see two files in the `route` folder—`index.js` and `users.js`. Those files hold information related to their routes, respectively. Here's the information regarding the route of `index.js`:

```
var express = require('express');
var router = express.Router();

/* GET home page. */
router.get('/', function(req, res, next) {
  res.render('index', { title: 'Express' });
});

module.exports = router;
```

Here's the information regarding the route of `users.js`:

```
var express = require('express');
var router = express.Router();

/* GET users listing. */
router.get('/', function(req, res, next) {
  res.send('respond with a resource');
});

module.exports = router;
```

In summary, routes are how an application's endpoints (URIs) respond to client requests:

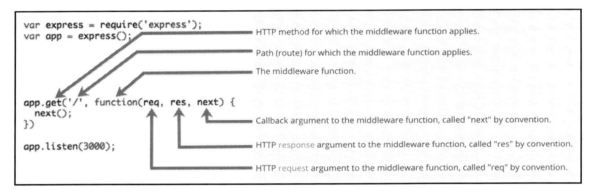

Express routes definition. Source: https://expressjs.com/en/guide/writing-middleware.html.

Finally, the `views` folder contains all the information about the view part of the application. Since we used `.hbs` to generate this dummy application, the handlebars files, that is, `.hbs` files, will be there.

 If you want to learn more about handlebars, go to their documentation at https://handlebarsjs.com/.

Now, it is time to see this dummy application in action. Like we did before, go to the root folder and install the necessary dependencies:

```
$ npm install
```

You can also use the shortened version:

```
$ npm i
```

Right after the installation, start the application:

```
$ npm start
```

Then, go to the browser under port 3000:

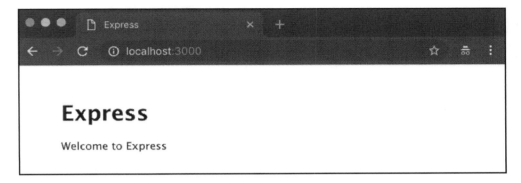

First Hello World with Express visible from the web browser

At the time of writing this book, Express.js is under development with Version 5. Version 5 is not very different from Version 4, but the differences can be seen at https://expressjs.com/en/guide/migrating-5.html.

If you want to use Version 5, the following command will create the skeleton for you:

```
$ npm install express@>=5.0.0-alpha.7
```

TypeScript installation

We're finally at the TypeScript installation stage. Now that we have all of the dependencies installed, TypeScript installation is a piece of cake through npm:

```
$ npm install -g typescript
```

And that's it. TypeScript is ready to be used.

To create our first TypeScript application, create a folder called my-first-ts-app and in this folder, create a file called hello.ts with the following content:

```
function hello(person) {
    return "Hello, " + person;
}
```

```
let user = "John User";

document.body.innerHTML = hello(user);
```

As you can see, it used a `.ts` extension for the `hello.ts` file, which means that this code is just JavaScript.

Now, it is time to compile the `hello.ts` file by running the following command at the command line:

$ tsc hello.ts

The result has to be a file called `hello.js`, which contains the same JavaScript that you filled in at the `hello.ts` file. So, if the code is the same, why should you use TypeScript?

Go to the `hello.ts` file and annotate the `string` type as a parameter in the `hello` function:

```
function hello(person: string) {
    return "Hello, " + person;
}

let user = "John User";

document.body.innerHTML = hello(user);
```

Now, it is definitely clear that the `hello` function expects a string as a parameter. This means that if we change the type of the `user` variable, the compilation will fail:

```
function hello(person: string) {
    return "Hello, " + person;
}

let user = 20;

document.body.innerHTML = hello(user);
```

Run the following command:

$ tsc hello.ts

You will see the following response:

```
hello.ts:7:33 - error TS2345: Argument of type 'number' is not assignable
to parameter of type 'string'.

7 document.body.innerHTML = hello(user);
```

In `Chapter 5`, *Building Your First API*, we will walk through TypeScript's features in detail.

Visual Studio Code

One of the greatest integrations with TypeScript is Visual Studio Code, which is also available from Microsoft. This section is going to guide you on its installation and integration with TypeScript. Visual Studio Code is available for Windows, Linux, and macOS. There are a lot of plugins and support, such as debugging, embedded Git control, syntax highlighting, intelligent code completion, snippets, code refactoring, and so on.

To get Visual Studio Code, the first step is to go to its website, `https://code.visualstudio.com/`. Then, under the **Download** section, download the version that you need based on your operating system:

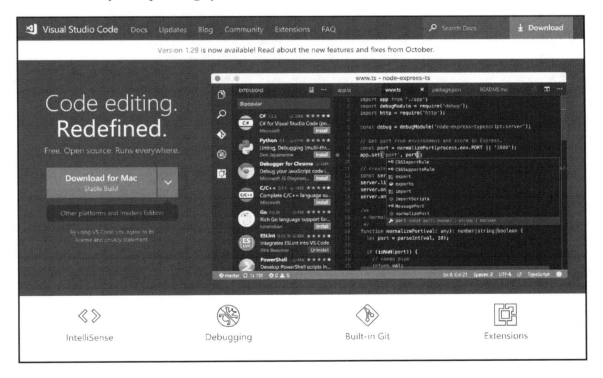

Visual Studio website

When the download is complete, double-click the file and follow the instructions to get Visual Studio Code installed:

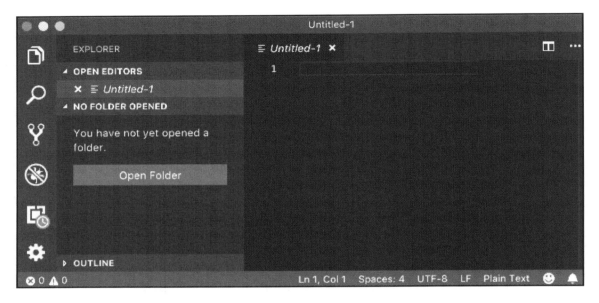

First impression of Visual Studio Code

Following the icons on the top left-hand side, there's the project explorer, search, source control, debug, and extensions, which you can use to add new languages, themes, debuggers, and so on before connecting them to additional services. One interesting thing is that the extensions process runs in processes that are separated from the rest so that they don't mess with the editor and slow it down.

Code Linters

In this section, we will learn how to get code Linters ready through Visual Studio Code. We will start with TSLint.

TSLint

When it comes Linters for TypeScript, the first one that comes to mind is TSLint. TSLint is an extensible static analysis tool, like so many Linters that check TypeScript code. Linters are used to enforce a strict and consistent code style in a project and to check for some simple yet common programming errors. This improves the readability and maintainability of the code.

TSLint is supported by VS Code and other modern editors and build systems. Another interesting feature is that TSLint can be customized with rules you might want to create.

Currently, TSList is available at `https://palantir.github.io/tslint/`, and the process we will follow to install it is really easy. You just have to go to your Terminal and use `npm`:

```
$ npm install tslint typescript -g
```

The output should be something similar to the following:

```
/usr/local/bin/tslint -> /usr/local/lib/node_modules/tslint/bin/tslint
/usr/local/bin/tsserver ->
/usr/local/lib/node_modules/typescript/bin/tsserver
/usr/local/bin/tsc -> /usr/local/lib/node_modules/typescript/bin/tsc
+ tslint@5.11.0
+ typescript@3.2.1
added 40 packages from 20 contributors and updated 1 package in 26.926s
```

To check whether TSLint was installed successfully, check its version:

```
$ tslint -v
```

You should be able to see the currently installed version as output.

Now, it is time to add a TSLint extension to the **Visual Studio Code (VS Code)** editor. For that, you just have to go to the extension options on Visual Studio Code and type in `tslint`. There will be dozens of different Linters to choose from. We decided to go with **TSLint** from **egamma**, which can be found at `https://github.com/Microsoft/vscode-tslint`:

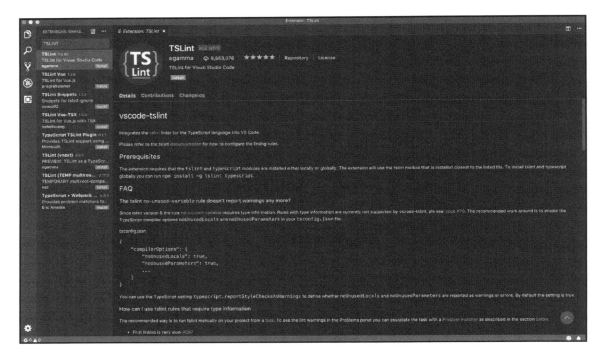

TSLint VS Code extension from egamma

To install it, click **Install** and wait for it to complete. When it finishes, click on **Reload to Activate**:

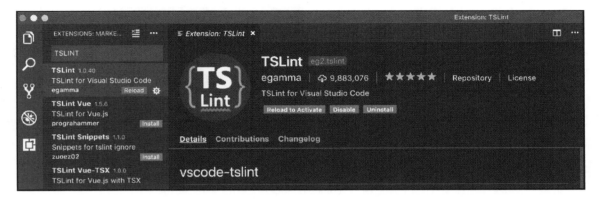

TSLint egamma installation page

Now, it is time to test the TSLint plugin. Create a new folder called `my-first-tslint-app` and run the following command:

```
$ tsc --init
```

When you open it with VS Code, you should see the following structure:

Content of the tsconfig.json configuration file

The only file you should see is the `tsconfig.json` file, with a few default values and an extensive list of commented-out possible configurations. The cleaner version of this file, without the commented ones, might look something like this:

```
{
  "compilerOptions": {
  "target": "es5",
  "module": "commonjs",
  "strict": true,
  "esModuleInterop": true
  }
}
```

We will also add two more there—outDir, which is going to say where compiled files go, and sourceMap, which is used to generate source maps:

```
{
  "compilerOptions": {
    "target": "es5",
    "module": "commonjs",
    "strict": true,
    "esModuleInterop": true,
    "outDir": "dist",
    "sourceMap": true
  }
}
```

After that, you can create an index.ts file by right-clicking on the explorer view:

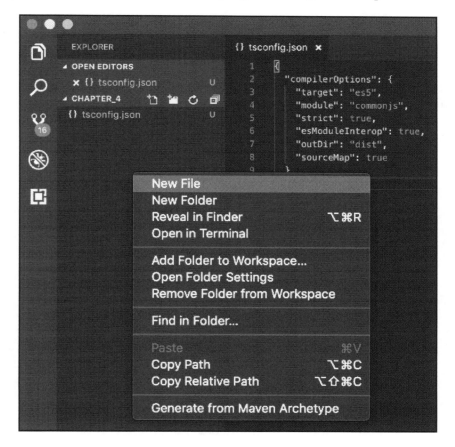

New folder creation in VS Code

It should have the following content:

```
const message = 'default message';

export function hello(word: string = message): string {
  return `Hello ${message}! `;
}
```

Now, go to the top navigation menu and click on **Terminal** | **New Terminal**:

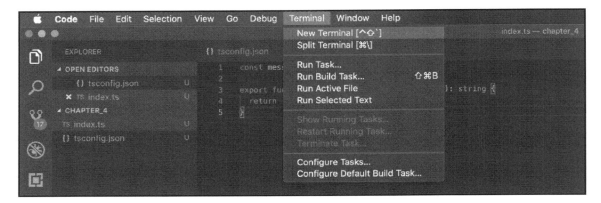

Menu option for integrated Terminal from VS Code

You will see the Terminal as shown:

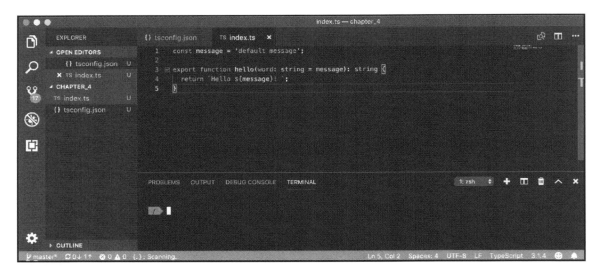

The integrated Terminal from VS Code

To compile the code, run the following command in your Terminal:

```
$ tsc
```

The output will be visible through the dist folder, that is, the index.js and index.js.map files:

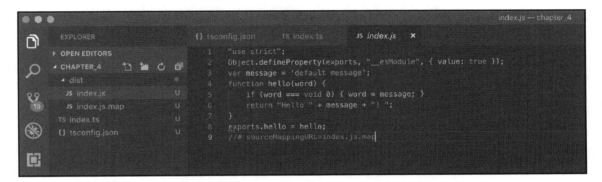

index.js file from VS Code

You can also set the tsc command to run every time that the code changes, also known as watch mode, rather than running the tsc command manually every time a change is made:

```
$ tsc -w
```

To get TSLint into this code, run the following command in the my-first-tslint-app folder:

```
$ tslint --init
```

This command will initialize TSLint for this application by adding a new file called tslint.json:

```
{
    "defaultSeverity": "error",
    "extends": [
        "tslint:recommended"
    ],
    "jsRules": {},
    "rules": {},
    "rulesDirectory": []
}
```

After that, if you go to the `index.ts` file again, you should see two messages from TSLint there:

- **[tslint] ' should be " [quotemark]:**

```
TS index.ts  ✕              [tslint] ' should be " [quotemark]
  1    const message = 'default message';
  2
  3    export function hello(word: string = message): string {
  4      return `Hello ${message}! `;
  5    }
```

- **[tslint] file should end with a newline [eofline]:**

```
TS index.ts  ✕
  1    const message = 'default message';
  2
  3    export function hello(word: string = message): string {
  4          [tslint] file should end with a newline [eofline]
  5    }
```

Let's say that we agree with the first message. Let's fix it by changing the single quotes to double quotes:

```
const message = "default message";

export function hello(word: string = message): string {
  return `Hello ${message}! `;
}
```

However, we don't agree with the second lint. To remove this validation, we can simply add a new rule to the `tslint.json` file, like so:

```
{
    "defaultSeverity": "error",
    "extends": [
        "tslint:recommended"
    ],
    "jsRules": {},
```

```
    "rules": {
        "eofline": false
    },
    "rulesDirectory": []
}
```

After you have saved the file, the message will disappear.

Currently, there are a lot of pre-generated Linters style available on the internet, which could be used if you configure rule by rule every time you want to create a new project. One of the most famous is the Airbnb style, and it is really easy to use it. Use the following command to do so:

```
$ npm install tslint-config-airbnb
```

From here, change the tslint.json configuration file so that it extends the Airbnb configuration instead:

```
{
    "defaultSeverity": "error",
    "extends": [
        "tslint-config-airbnb",
    ],
    "jsRules": {},
    "rules": {
        "eofline": false
    },
    "rulesDirectory": []
}
```

To make our lives even easier, we can use gts, which is Google TypeScript style guide, and the configuration for our formatter, Linter, an automatic code fixer.

Let's test it out by creating a new folder called my-first-gts-project. In this folder, type:

```
$ npx gts init
```

The preceding command will generate everything you need to get started with a tsconfig.json file and Linting setup. This command will also generate the project's package.json file:

```
npx: installed 180 in 21.875s
package.json does not exist.
? Generate Yes
Writing package.json...
{ scripts:
```

```
    { test: 'echo "Error: no test specified" && exit 1',
      check: 'gts check',
      clean: 'gts clean',
      compile: 'tsc -p .',    .
      fix: 'gts fix',
      prepare: 'npm run compile',
      pretest: 'npm run compile',
      posttest: 'npm run check' },
  devDependencies: { gts: '^0.9.0', typescript: '~3.1.0' } }
Writing ./tsconfig.json...
{
  "extends": "./node_modules/gts/tsconfig-google.json",
  "compilerOptions": {
    "rootDir": ".",
    "outDir": "build"
  },
  "include": [
    "src/*.ts",
    "src/**/*.ts",
    "test/*.ts",
    "test/**/*.ts"
  ]
}

Writing ./tslint.json...
{
  "extends": "gts/tslint.json"
}

Writing ./.clang-format...
Language: JavaScript
BasedOnStyle: Google
ColumnLimit: 80
npm notice created a lockfile as package-lock.json. You should commit this
file.
npm WARN @0.0.0 No description
npm WARN @0.0.0 No repository field.

added 181 packages from 84 contributors and audited 306 packages in 5.878s
found 0 vulnerabilities
```

The default skeleton is as shown:

```
package.json — my-first-gts-project

EXPLORER                    {} package.json ×

▲ OPEN EDITORS               1  {
  × {} package.json    2, U   2      "name": "",
▲ MY-FIRST-GTS-PROJECT        3      "version": "0.0.0",
  ▶ node_modules              4      "description": "",
  ≡ .clang-format        U    5      "main": "build/src/index.js",
  {} package-lock.json        6      "types": "build/src/index.d.ts",
  {} package.json       2, U  7      "files": [
  {} tsconfig.json       U    8          "build/src"
  {} tslint.json         U    9      ],
                             10      "license": "Apache-2.0",
                             11      "keywords": [],
                             12      "scripts": {
                             13          "test": "echo \"Error: no test specified\" && exit 1",
                             14          "check": "gts check",
                             15          "clean": "gts clean",
                             16          "compile": "tsc -p .",
                             17          "fix": "gts fix",
                             18          "prepare": "npm run compile",
                             19          "pretest": "npm run compile",
                             20          "posttest": "npm run check"
                             21      },
                             22      "devDependencies": {
                             23          "gts": "^0.9.0",
                             24          "typescript": "~3.1.0"
                             25      }
                             26  }
                             27

▶ OUTLINE
```

Default gts skeleton

After that, create a folder under the root project folder called `src` and a file called `index.ts`:

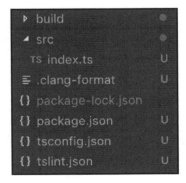

Default gts resources

Now, you can simply install the dependencies:

```
$ npm install
```

And we're done! We've just gone through an easy way to get started with TypeScript using TSLint configurations. At the time of writing this book, the `gts` framework was under development with a non-stable version.

Prettier

Prettier is an interesting tool because, every time you save the file, the code is formatted. It also integrates with most editors, such as VS Code.

To get Prettier working with our project, let's install it. Go into the `my-first-gts-project` folder and run the following command:

```
$ npm install prettier --save-dev --save-exact
```

After the installation is complete, go to the `index.ts` file and put all of the following code in a single line:

```
const message = "default message"; export function hello(word: string =
message): string {return `Hello ${message}! `;}
```

Then, check it by running `prettier` against this file:

```
$ npx prettier --write src/index.js
```

You should see the code formatted again:

```
const message = "default message";
export function hello(word: string = message): string {
  return `Hello ${message}! `;
}
```

You can also install the Prettier extension on VS Code:

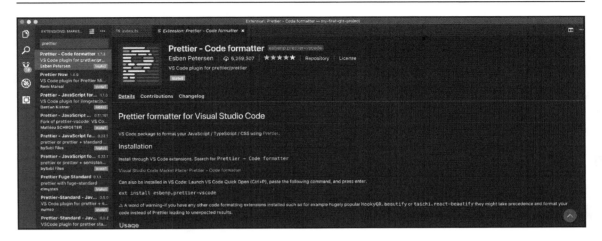

Prettier installation page from VS Code

To enable the autosaving feature of Prettier, you have to go to **VSCode configuration Code | Preferences | Settings**, select **Open settings.json**, and add the property:

file.json settings from VS Code

Then, add the following code there:

```
"editor.formatOnSave": true
```

Testing suites

There are a lot of books that only cover testing strategies such as **Test-Driven Design** (TDD), **Behavior-Driven Design** (BDD), unit tests, integration tests, smoke tests, regression tests, and so on. The scope of this book allows us to create a minimal but useful testing suite so that you can start based on this model and improve on it, depending on your project's needs. Basically, this chapter covers the following essentials:

- Providing a testing structure with Mocha
- Using assertion functions with Chai
- Mutation tests with Stryker

This chapter won't show you how to implement all of these, since they will be covered later in the book. At the moment, our focus is on installing all of the necessary dependencies while explaining how they work at a high level.

For the purpose of testing suites, we will create a folder called `my-first-suite-test-project`, which we will use throughout this section. The command to create this new folder is as follows:

```
$ mkdir my-first-suite-test-project
```

Go to the `my-first-suite-test-project`and folder and initialize a new project:

```
$ npm init -y
```

A `package.json` file will be created. Run the following command to get `ts-node` and TypeScript:

```
$ npm install --save-dev ts-node typescript
```

We also have to initialize TypeScript with the following command:

```
$ tsc --init
```

The `tsconfig.json` file may look as follows:

```
{
  "compilerOptions": {
    "target": "es5",
    "module": "commonjs",
    "strict": true,
    "esModuleInterop": true
  }
}
```

The following sections will guide you through how to install Mocha and Chai.

Mocha and Chai

Mocha is a JavaScript test framework for Node.js applications. Mocha makes asynchronous testing simple and runs serially.

To get Mocha, you can use npm to install it:

```
// global installation
$ npm install --g mocha
```

Also install mocha types:

```
// local installation
$ npm install --save-dev mocha @types/mocha
```

And that's it. Mocha is already installed and ready to be used. The next step is to install Chai.

 Like the other frameworks, it is recommended that you go through the Mocha documentation, which is available at https://mochajs.org/api/, to understand it further.

Chai is a BDD and TDD assertion framework that was designed for Node.js applications.

 Chai's documentation is available at https://www.chaijs.com/api/.

The installation for Chai is also simple and is available through npm:

```
// global installation
$ npm install --g chai

// local installation
$ npm install --save-dev chai @types/chai
```

Simple! Now that we have Mocha and Chai installed, it is time to start writing some tests.

Create a folder called `src` and a file called `Queue.ts` inside of that folder with the following content:

```
export class Queue<T> {

    private _store: T[] = [];

    constructor(initialData: Array<T> = []) {
        this._store.push(...initialData)
    }

    push(val: T) {
      this._store.push(val);
    }

    pop(): T | undefined {
      return this._store.shift();
    }

    isEmpty(): boolean {
        return this.size() === 0;
    }

    size(): number {
        return this._store.length;
    }

}
```

This is a classical queue FIFO implementation, which we will use to run some tests against it. After that, create a new file on the same level as `Queue.ts`, called `app.ts`, with the following content:

```
export { Queue } from './Queue';
```

In the `package.json` file, change the `script` property so that it has the following information:

```
{
  "name": "my-first-suite-test-project",
  "version": "1.0.0",
  "description": "",
  "main": "src/app.js",
  "scripts": {
    "test": "mocha --require ts-node/register test/**/*.spec.ts"
  },
  "keywords": [],
  "author": "",
```

```
    "license": "ISC",
    "devDependencies": {
        "@types/chai": "^4.1.7",
        "@types/mocha": "^5.2.5",
        "chai": "^4.2.0",
        "mocha": "^5.2.0",
        "ts-node": "^7.0.1",
        "typescript": "^3.2.1"
    }
}
```

Under the "test" property, we are telling the script that we want to use Mocha. --require ts-node/register registers TypeScript as on-the-fly transpilers so that we can write the tests in TypeScript. Also, it means that our tests will have .spec.ts in the filenames.

Now, it is time to write some tests. Create a new folder called test and, inside that folder, create a new file called queue.spec.ts with the following content:

```
import { expect } from 'chai';
import { Queue } from '../src/app';

describe('Queue', () => {
    it('should be able to be initialized without an initializer', () => {
        const s = new Queue<number>();
        expect(s.size()).to.equal(0);
    });
    it('should be able to be initialized with an initializer', () => {
        const s = new Queue<number>([ 1, 2, 3, 4 ]);
        expect(s.size()).to.equal(4);
    });
    it('should be able to add a new element after initialized', () => {
        const s = new Queue<number>([ 1, 2, 3, 4 ]);
        s.push(5);
        expect(s.size()).to.equal(5);
        expect(s.pop()).to.equal(1);
    });
    it('should be able to get the first element', () => {
        const s = new Queue<number>([ 1, 2, 3, 4 ]);
        expect(s.pop()).to.equal(1);
    });
});
```

Finally, let's run some tests:

```
$ npm run test
```

The output might be something like this:

```
> mocha --require ts-node/register test/**/*.spec.ts


  Stack
    should be able to be initialized without an initializer
    should be able to be initialized with an initializer
    should be able to add a new element after initialized
    should be able to get the first element


  4 passing (16ms)
```

Stryker

Stryker is a mutation testing tool. Mutation tests are used to provide a new mindset when designing tests. It evaluates the quality of existing software tests by modifying a program in small batches. The main idea is mutating the behavior of the tests while detecting and rejecting mutations that might happen.

Let's say you are writing some code to validate whether a kid can ride a roller coaster:

```
function isAllowedToRideARollerCoaster(kid) {
  return kid.height >= 318.3;
}
```

Stryker will find the `return` statement and decide to change it in a few different ways, like so:

```
/* 1 */ return kid.height > 360;
/* 2 */ return kid.height < 250;
/* 3 */ return false;
/* 4 */ return true;
```

Stryker calls those modifications mutants. So, for the preceding example, there are four mutants, where two of them will pass and two will fail. Stryker applies them one by one and your tests are executed against them. The mutant is classified as killed if at least one of the tests fails.

To get Stryker, we need to use npm to install it. First, let's install the Stryker command-line interface:

```
$ npm install -g stryker-cli
```

Now, let's install Stryker on the project:

```
$ npm install --save-dev @stryker-mutator/core stryker-api
```

To configure Stryker, we need to `init` it:

```
$ stryker init
```

After you run this command, an interactive prompt will appear with some questions.
Follow the following process:

```
? Are you using one of these frameworks? Then select a preset
configuration. None/other
? Which test runner do you want to use? If your test runner isn't listed
here, you can choose "command" (it uses your `npm test` command, but will
come with a big performance penalty) mocha
? Which test framework do you want to use? mocha
? What kind of code do you want to mutate? typescript
? [optional] What kind transformations should be applied to your code?
typescript
? Which reporter(s) do you want to use? html, clear-text, progress
? Which package manager do you want to use? npm
Writing stryker.conf.js...
Installing NPM dependencies...
npm i --save-dev stryker-api stryker-mocha-runner stryker-mocha-framework
stryker-typescript stryker-typescript stryker-html-reporter
npm WARN my-first-gts-project@0.0.0 No description
npm WARN my-first-gts-project@0.0.0 No repository field.

+ stryker-api@0.22.0
+ stryker-html-reporter@0.16.9
+ stryker-mocha-runner@0.15.2
+ stryker-mocha-framework@0.13.2
+ stryker-typescript@0.16.1
+ stryker-typescript@0.16.1
added 16 packages from 11 contributors, updated 1 package and audited 636
packages in 7.39s
found 0 vulnerabilities

Done configuring stryker. Please review `stryker.conf.js`, you might need
to configure transpilers or your test runner correctly.
Let's kill some mutants with this command: `stryker run`
```

After you finish prompting the Stryker information, a file called `stryker.conf.js` will appear with the following information:

```
module.exports = function(config) {
  config.set({
    mutator: "typescript",
    packageManager: "npm",
    reporters: ["html", "clear-text", "progress"],
    testRunner: "mocha",
    transpilers: ["typescript"],
    testFramework: "mocha",
    coverageAnalysis: "perTest",
    tsconfigFile: "tsconfig.json",
    mutate: ["src/**/*.ts"]
  });
};
```

Now, we are able to write our first mutation code by running the following command:

$ stryker run

This command may fail outside a Git repository. If you face this error, run the `git init` command and then set the coverage analysis attribute to `"of"` in the `stryker.conf.js` file.

You should be able to see the following output:

```
00:14:58 (49306) INFO ConfigReader Using stryker.conf.js in the current
working directory.
00:14:58 (49306) INFO TypescriptConfigEditor Loading tsconfig file
/Users/biharck/Developer/Hands-On-RESTful-Web-Services-with-
TypeScript-3/Chapter04/my-first-suite-test-project/tsconfig.json
00:14:59 (49306) INFO InputFileResolver Found 2 of 9 file(s) to be mutated.
00:14:59 (49306) INFO InitialTestExecutor Starting initial test run. This
may take a while.
00:15:00 (49306) INFO InitialTestExecutor Initial test run succeeded. Ran 4
tests in 1 second (net 6 ms, overhead 50 ms).
00:15:00 (49306) INFO Stryker 10 Mutant(s) generated
00:15:01 (49306) INFO SandboxPool Creating 3 test runners (based on CPU
count)
Mutation testing [====================================================] 100%
(ETC n/a) 7/7 tested (0 survived)

4. [NoCoverage] BinaryExpression
/Users/biharck/Developer/Hands-On-RESTful-Web-Services-with-
TypeScript-3/Chapter04/my-first-suite-test-project/src/Queue.ts:17:34
- return this._store.length === 0;
```

```
+ return this._store.length !== 0;

5. [NoCoverage] ConditionalExpression
/Users/biharck/Developer/Hands-On-RESTful-Web-Services-with-
TypeScript-3/Chapter04/my-first-suite-test-project/src/Queue.ts:17:15
- return this._store.length === 0;
+ return false;

6. [NoCoverage] ConditionalExpression
/Users/biharck/Developer/Hands-On-RESTful-Web-Services-with-
TypeScript-3/Chapter04/my-first-suite-test-project/src/Queue.ts:17:15
- return this._store.length === 0;
+ return true;

Ran 0.50 tests per mutant on average.
----------|----------|-----------|-----------|------------|----------|-------
--|
File | % score | # killed | # timeout | # survived | # no cov | # error |
----------|----------|-----------|-----------|------------|----------|-------
--|
All files | 40.00 | 2 | 0 | 0 | 3 | 5 |
 Queue.ts | 40.00 | 2 | 0 | 0 | 3 | 5 |
----------|----------|-----------|-----------|------------|----------|-------
--|
00:15:03 (49306) INFO HtmlReporter Your report can be found at:
file:///Users/biharck/Developer/Hands-On-RESTful-Web-Services-with-
TypeScript-3/Chapter04/my-first-suite-test-
project/reports/mutation/html/index.html
00:15:03 (49306) INFO Stryker Done in 4 seconds.
```

An HTML report like the following will be available, and is generated in the
reports folder:

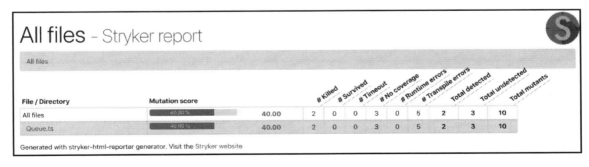

Main report page from Stryker

Here, we can see that the **Mutation score** is **40%** and that **2** mutations were killed. By clicking on the `Queue.ts` link, we can see that the method is empty and was not covered by tests:

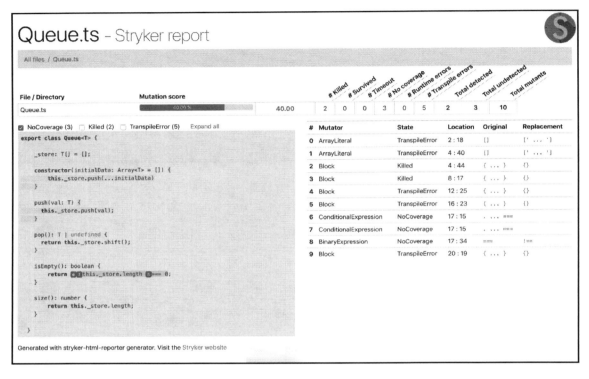

Detailed Stryker report

To cover this method, let's write a new test to check whether the queue is empty:

```
import { expect } from 'chai';
import { Queue } from '../src/app';

describe('Queue', () => {
    it('should be able to be initialized without an initializer', () => {
        const s = new Queue<number>();
        expect(s.size()).to.equal(0);
    });
    it('should be able to be initialized with an initializer', () => {
        const s = new Queue<number>([ 1, 2, 3, 4 ]);
        expect(s.size()).to.equal(4);
    });
    it('should be able to add a new element after initialized', () => {
        const s = new Queue<number>([ 1, 2, 3, 4 ]);
        s.push(5);
        expect(s.size()).to.equal(5);
        expect(s.pop()).to.equal(1);
    });
    it('should be able to get the first element', () => {
        const s = new Queue<number>([ 1, 2, 3, 4 ]);
        expect(s.pop()).to.equal(1);
    });
    it('should check if the queue is empty when there is no data there', ()
=> {
        const s = new Queue<number>([ ]);
        expect(s.isEmpty()).to.equal(true);
    });
});
```

So, as you can see, the results are a little bit better, with an **80%** of **Mutation score**:

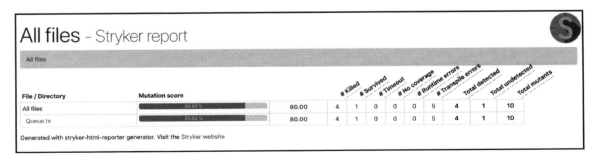

If we analyze what we are still missing, we will see that there is a mutation that survived:

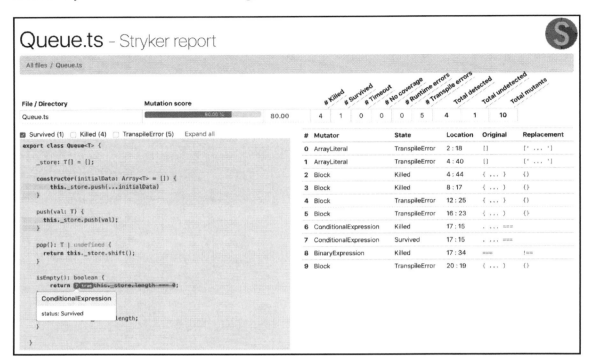

Stryker changed the `this._store.length` expression to `true`, but the test still passed. To fix this issue, we could change the `this._store.length` expression to use the `size()` method:

```
isEmpty(): boolean {
  return this.size() === 0;
}
```

By running Stryker again, the result will be 100%:

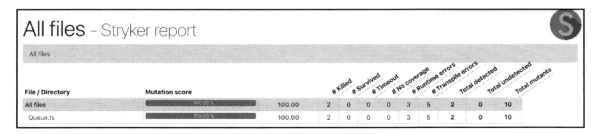

Debugging

One of the features that modern editors provide is the ability to debug code interactively, which allows the developer to be more productive and find out what happens when code fails and when troubleshooting occurs.

This section is going to walk you through how to enable debugging using VS Code. Don't worry—this is rather easy. Just follow the steps as shown:

1. To do so, click on the **Debug** item on the left-hand side of the VS Code toolbar:

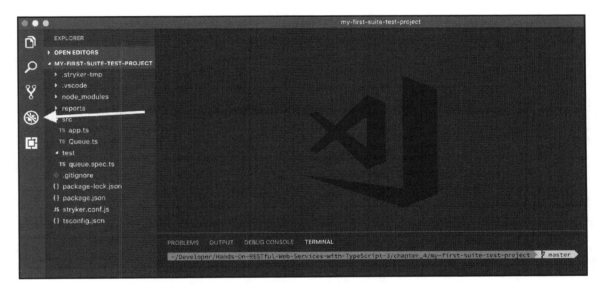

Debugging option from VS Code

2. Then, click on the drop-down menu and select **Add Configuration**:

Debug configuration section

3. Select **() Node.js: Mocha Tests**:

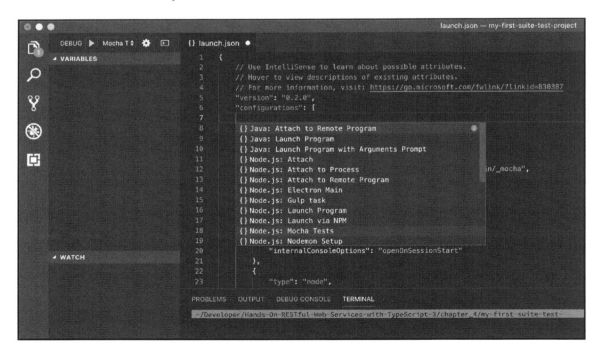

Debugging configuration file

After you click on **() Node.js: Mocha Tests**, a new configuration will be written under the `launch.json` file with the following information:

```
{
    "version": "0.2.0",
    "configurations": [
        {
            "type": "node",
            "request": "launch",
            "name": "Mocha Tests",
            "program":
"${workspaceFolder}/node_modules/mocha/bin/_mocha",
            "args": [
                "--require", "ts-node/register",
                "-u", "bdd",
                "--timeout", "999999",
                "--colors", "--recursive",
                "${workspaceFolder}/test/**/*.ts"
            ],
            "internalConsoleOptions": "openOnSessionStart"
        }
    ]
}
```

4. Now, the next step is to add a breakpoint and start debugging, like so:

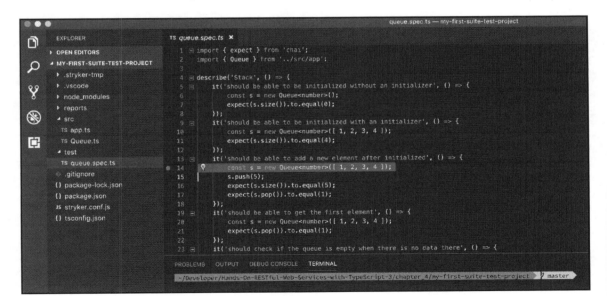

Adding a breakpoint into a test file

5. The next step is to start the debug session by clicking on the start icon, which is shown in the following screenshot:

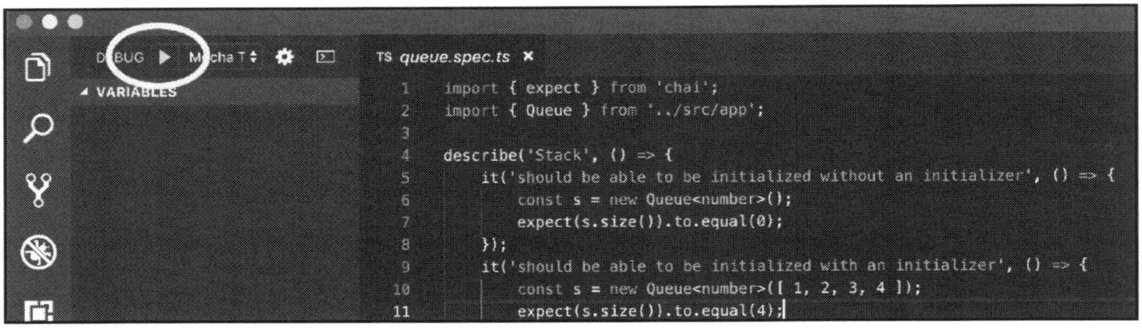

The start icon under the debug section

Since the debugging process has already started, you can use the debugging tool at the top to walk through the code, stop the process, and so on:

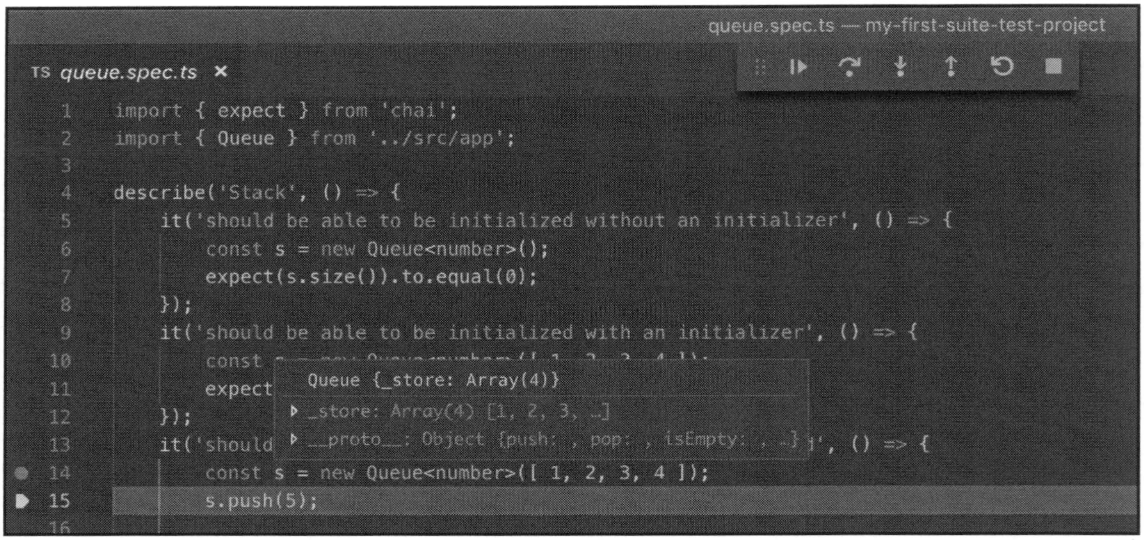

Debugging in action

Building tasks for coding

During the development process, we may repeat a lot of processes that we could automate or at least group them into tasks to help ourselves out. This section is going to show you some ideas for tasks that might help you not spend so much time on a repeated process.

If you press ⇧⌘B (or *Ctrl + Shift + B* on Windows/Linux) from the global Terminal menu, you will see the build tasks that are available:

Global build tasks from VS Code

Selecting `tsc:build - tsconfig.json` will produce the `.js` files in the `dist` folder, as well as the `.js.map` files that will be created next to the `.ts` files since the `outDir`.

Another task that's available is `tsc:watch tsconfig.json`, which makes the TypeScript compiler watch for changes to your TypeScript files and runs the transpiler on each change.

There is also the possibility to define the TypeScript build task as the default build task. By doing so, every time you run the build task (⇧⌘B), the default task will be executed. To do so, select **Configure Default Build Task** and select **TypeScript tsc : build**. The following information should be included in the `tasks.json` file:

```
{
    // See https://go.microsoft.com/fwlink/?LinkId=733558
    // for the documentation about the tasks.json format
    "version": "2.0.0",
    "tasks": [
        {
            "type": "typescript",
            "tsconfig": "tsconfig.json",
            "problemMatcher": [
                "$tsc"
            ],
```

```
            "group": {
                "kind": "build",
                "isDefault": true
            }
        }
    ]
}
```

Summary

In this chapter, we have talked about starting from scratch with TypeScript by passing through the installation and configuration of Node.js. We also presented the Express.js server using Node.js and our first `hello world` application. To get all of the necessary tools installed, we talked about `npm` and how powerful this tool is during the development process.

After we installed Node.js, we walked through TypeScript's installation and preparing the development environment by using VS Code and the basics of TypeScript, such as configuring Linters to help us with patterns and formatting your coding environment for better coding.

The last part of this chapter covered testing tools, different types of tests, how to set up and debug using VS Code, and how to define build tasks for transpiling TypeScript code and building the application.

The next chapter will walk you through building out your first API.

Questions

1. Is it possible to use Express without Node.js?
2. What does NPM do?
3. What does it mean to route in Express?
4. What is VS Code?
5. What does Linters do?
6. What do you need to do to generate source maps?
7. What does Mocha do?
8. What does Chai do?
9. What does Stryker do?

Further reading

To improve your knowledge of what was covered in this chapter, the following books are recommended, as they will be helpful for the upcoming chapters:

- *Beginning API Development with Node.js* (`https://www.packtpub.com/web-development/beginning-api-development-nodejs`)
- *RESTful Web API Design with Node.js* (`https://www.packtpub.com/web-development/restful-web-api-design-nodejs-10-third-edition`)
- *Using Node.js UI Testing* (`https://www.packtpub.com/web-development/using-nodejs-ui-testing`)
- *Node.js Essentials* (`https://www.packtpub.com/web-development/nodejs-essentials`)
- *Mastering Typescript 3* (`https://www.oreilly.com/library/view/mastering-typescript-3/9781789536706/`)

5
Building Your First API

Your first API, *Hello World*, will mainly focus on how to start the app that is going to serve as your web service. This chapter will describe file organization and folder structures so that we can create a more maintainable and scalable code base. Then, we will focus on how to define routes with a very classic *Hello World* output as a result of a web service call. Finally, we will show you the controller logic that is going to run when a certain endpoint is called.

Throughout this book, we are going to create a simple **Order Management System (OMS)**, which will connect with a NoSQL database. The main output of this service is to expose a REST API. This chapter will start this engagement from a basic *Hello World* service, which will reshape itself accordingly.

The following topics will be covered in this chapter:

- Serving the app
- File structure
- Defining basic routes
- Controlling basic routes

Technical requirements

All of the information that's required to run the code in this chapter is provided in the necessary sections. The only requirement is that you complete all of the installations from the previous chapter, that is, for Node.js, VS Code, TypeScript, and so on.

All of the code that's used in this chapter is available at `https://github.com/PacktPublishing/Hands-On-RESTful-Web-Services-with-TypeScript-3/tree/master/Chapter05`.

Serving the app

Based on the previous chapters, we will walk you through the initial configuration of an Order API application based on the Swagger application we created in Chapter 3, *Designing RESTful APIs with OpenAPI and Swagger*, and the inputs we looked at in Chapter 4, *Setting Up Your Development Environment*.

This section will guide you on the project's basic configuration, such as the package dependencies, testing tools, Linters, and so on. After that, the next step is to start coding using some **Test-Driven Development (TDD)** principles. Even though this book is not based on TDD principles, it is definitely a good practice to start coding from testing.

Initial configurations and file structure

Since we know how a TypeScript application works, we can start with the basic configurations in place. To start our application development, let's create a folder called order-api and create a file called package.json inside it with the following information:

```
{
  "name": "order-api",
  "version": "1.0.0",
  "description": "This is the example from the Book Hands on RESTful Web Services with TypeScript 3",
  "main": "./dist/server.js",
  "scripts": {
    "build": "npm run clean && tsc",
    "clean": "rimraf dist && rimraf reports",
    "lint": "tslint ./src/**/*.ts ./test/**/*.spec.ts",
    "lint:fix": "tslint --fix ./src/**/*.ts ./test/**/*.spec.ts -t verbose",
    "pretest": "cross-env NODE_ENV=test npm run build && npm run lint",
    "test": "cross-env NODE_ENV=test mocha --reporter spec --compilers ts:ts-node/register test/**/*.spec.ts ",
    "test:mutation": "stryker run",
    "stryker:init": "stryker init",
    "dev": "cross-env PORT=3000 NODE_ENV=dev ts-node ./src/server.ts",
    "prod": "PORT=3000 npm run build && npm run start",
    "tsc": "tsc"
  },
  "engines": {
    "node": ">=8.0.0"
  },
  "keywords": [
```

```
      "order POC",
      "Hands on RESTful Web Services with TypeScript 3",
      "TypeScript 3",
      "Packt Books"
  ],
  "author": "Biharck Muniz Araújo",
  "license": "MIT",
  "devDependencies": {
    "@types/body-parser": "^1.17.0",
    "@types/chai": "^4.1.7",
    "@types/chai-http": "^3.0.5",
    "@types/express": "^4.16.0",
    "@types/mocha": "^5.2.5",
    "@types/node": "^10.12.12",
    "chai": "^4.2.0",
    "cross-env": "^5.2.0",
    "mocha": "^5.2.0",
    "rimraf": "^2.6.2",
    "stryker": "^0.33.1",
    "stryker-api": "^0.22.0",
    "stryker-html-reporter": "^0.16.9",
    "stryker-mocha-framework": "^0.13.2",
    "stryker-mocha-runner": "^0.15.2",
    "stryker-typescript": "^0.16.1",
    "ts-node": "^7.0.1",
    "tslint": "^5.11.0",
    "tslint-config-prettier": "^1.17.0",
    "typescript": "^3.2.1"
  },
  "dependencies": {
    "body-parser": "^1.18.3",
    "express": "^4.16.4",
    "chai-http": "^4.2.1"
  }
}
```

As we already know, this file is responsible for maintaining the package's dependencies for all of the libraries that this project is going to use. Up until now, all of the dependencies have been from the previous chapters.

The next step is to add a file called `tsconfig.json`, which is responsible for keeping all TypeScript configuration together:

```
{
  "compilerOptions": {
    "declaration": true,
    "experimentalDecorators": true,
    "emitDecoratorMetadata": true,
```

```
      "lib": ["es6", "dom"],
      "target": "es6", //default is es5
      "module": "commonjs", //CommonJs style module in output
      "outDir": "dist", //change the output directory
      "resolveJsonModule": true //to import out json database
    },
    "include": [
      "src/**/*.ts" //which kind of files to compile
    ],
    "exclude": [
      "node_modules" //which files or directories to ignore
    ]
}
```

Now that we have the application package dependencies and TypeScript configuration in place, it is time to add two more files. One is responsible for defining the rules for the Linters. It is called `tslint.json` and includes the following data:

```
{
  "extends": ["tslint:recommended", "tslint-config-prettier"],
  "rules": {
    "array-type": [true, "generic"],
    "no-string-literal": false,
    "object-literal-shorthand": [true, "never"],
    "only-arrow-functions": true,
    "interface-name": false,
    "max-classes-per-file": false,
    "no-var-requires": false,
    "ban-types": false
  }
}
```

The second file is called `stryker.conf.js` and includes the configuration details for Stryker:

```
module.exports = function(config) {
  config.set({
    testRunner: 'mocha',
    mochaOptions: {
      files: ['test/**/*.spec.ts'],
      opts: './test/mocha.opts',
      ui: 'bdd',
      timeout: 35000,
      require: ['ts-node/register', 'source-map-support/register'],
      asyncOnly: false,
    },
    mutator: 'typescript',
    transpilers: [],
```

```
    reporters: ['html', 'progress', 'dashboard'],
    packageManager: 'npm',
    testFramework: 'mocha',
    coverageAnalysis: 'off',
    tsconfigFile: 'tsconfig.json',
    mutate: ['src/**/*.ts'],
  })
}
```

Finally, the latest configuration file is a file called `.prettierrc`, which is going to configure Prettier:

```
{
  "semi": false,
  "singleQuote": true,
  "trailingComma": "es5"
}
```

By the end of this, your application skeleton should be similar to the following:

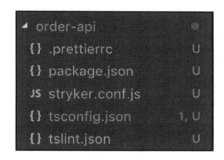

Base application skeleton

Now, it is time to install the necessary dependencies with the following command:

```
$ npm install
```

After using the preceding command, a folder called `node_modules` will appear on the same level as the configuration files, and you will have the project environment.

Defining and controlling basic routes

As we mentioned previously, we will start this project by writing tests. To do so, let's create a folder called `test` and, in this folder, create a file called `mocha.opts` with the following content:

```
--require ts-node/register
--require source-map-support/register
--full-trace
--bail
--timeout 35000
test/**/*.spec.ts
```

Those properties are going to configure Mocha with options such as `timeout`, the kind of files that will be used during the tests, and so on.

After that, create a folder called `routes` and, in this folder, create a file called `index.spec.ts` with the following content:

```
import * as chai from 'chai'
import chaiHttp = require('chai-http')
import 'mocha'
import app from '../../src/app'

chai.use(chaiHttp)

const expect = chai.expect

describe('baseRoute', () => {
  it('should respond with HTTP 200 status', async () => {
    return chai
      .request(app)
      .get('/index')
      .then(res => {
        expect(res.status).to.be.equal(200)
      })
  })
  it('should respond with success message', async () => {
    return chai
      .request(app)
      .get('/index')
      .then(res => {
        expect(res.body.status).to.be.equal('success')
      })
  })
})
```

The idea is that, even though there is no working code, we will write the tests based on the needs of this project, such as the following:

- When I hit the index endpoint—I get an HTTP status code of **200**:

```
it('should respond with HTTP 200 status', async () => {
  return chai
    .request(app)
    .get('/index')
    .then(res => {
      expect(res.status).to.be.equal(200)
    })
})
```

- The same applies when I hit the index endpoint—I get a message telling me that the process has succeeded:

```
it('should respond with success message', async () => {
  return chai
    .request(app)
    .get('/index')
    .then(res => {
      expect(res.body.status).to.be.equal('success')
    })
})
```

Now, create a folder on the same level as `test`, called `src`. In this folder, create a file called `app.ts` without anything inside of it.

Now, go to the Terminal and run the `clean`, `build`, and `test` commands, like so:

```
$ npm run clean
$ npm run build
$ npm run pretest
$ npm run test
```

As expected, some of the tests failed. So, now we can go ahead, write our first route, and serve the application.

The first step in serving the application is to write a class for the application's configuration. To do that, let's update the `app.ts` file with the following content:

```
import * as bodyParser from 'body-parser'
import * as express from 'express'

class App {
  public app: express.Application
```

```
  constructor() {
    this.app = express()
    this.app.use(bodyParser.json())
  }
}

export default new App().app
```

The first import is a tool that is used to parse the data that you pass in as the request body. Then, we import the express library.

Right after the imports, we will create a class called App. This class only does the basics, such as declaring an Express.js application. Now that we have the app.ts class already filled up, we are able to create the server.

The server is basically a file called server.ts that's on the same level as app.ts with the following content:

```
import app from './app'
const PORT = process.env.PORT

app.listen(PORT)
```

This code basically includes the App class's information and sets up the HTTP port where the application is going to become available.

Now, it is time to write our first route, called index.ts, in the order-api/src/routes/index.ts folder.

We will use the following code base for this:

```
import { Request, Response } from 'express'

export class Index {
  public routes(app): void {
    app.route('/index').get((req: Request, res: Response) => {
      res.status(200).send({ status: 'success' })
    })
  }
}
```

This route is a dummy route, and is being used to test our basic configuration. Once we have the index.ts file, we can go to the app.ts file an add the reference to the index.ts route:

```
import * as bodyParser from 'body-parser'
import * as express from 'express'
```

```
import { Index } from '../src/routes/index'

class App {
  public app: express.Application
  public indexRoutes: Index = new Index()

  constructor() {
    this.app = express()
    this.app.use(bodyParser.json())
    this.indexRoutes.routes(this.app)
  }
}

export default new App().app
```

Finally, we are now good to run the tests with the following command:

```
$ npm run test
```

We should see the following output:

```
baseRoute
  should respond with HTTP 200 status (75ms)
  should respond with success message


2 passing (88ms)
```

We can also run the Stryker `test` with the following command:

```
$ npm run test:mutation
```

The output will look something like this:

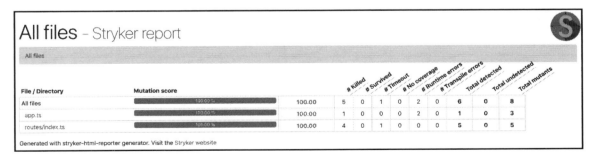

If you want to see the output from a browser, start the application:

```
$ npm run dev
```

Then, go to the browser and type in the following:

```
http://localhost:3000/index
```

The output will be a success message, like the following one:

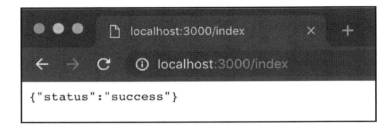

Hitting the initial project with a browser

Now, we have the initial configuration in place with a basic dummy routing of a GET operation, which returns a message with a status of success.

Testing with Postman

Postman is an application that allows developers to perform HTTP requests from a simple and intuitive interface, making it easy to test and debug REST services.

Postman is available as an application for the Google Chrome browser and has several useful features for developing this type of project. We can use it to test the actions of an ASP.NET Web API service by sending and receiving data in JSON format.

The documentation for Postman is available at https://learning.getpostman.com/docs/postman/.

The installation is really simple—you just have to download a version based on your operating system and follow the installation instructions that appear onscreen. The installation file is available at https://www.getpostman.com.

Once the installation has completed, you can launch Postman and use it to test endpoints, like so:

Consuming a REST endpoint with Postman

Summary

In this chapter, you were able to put everything that you have learned so far into practice in a meaningful way by creating the skeleton of the application we will use in this book. With that in place, we are now ready to start implementing the business logic and evolve the application.

You also gained an understanding of how to set up the Express.js server for your API, how to organize your files in a proper way, how to create your first route, and how to create controllers for your routes.

In the next chapter, we will walk you though enriching this base application by putting other routes in place. We will also explain how you can handle requests and responses.

Questions

1. Why do you need a `package.json` file?
2. What should you do if you want to use an external Linter rule?
3. Why is it suggested that you start coding from tests?
4. What is Postman?
5. Where should you add the proper HTTP status code?
6. What is meant by `dev` dependencies in the `package.json` file?
7. What does the `Mocha.opts` file do?

Further reading

To improve your knowledge of what was covered in this chapter, the following books are recommended, as they will be helpful for the upcoming chapters:

- *RESTful Web API Design with Node.js* (`https://www.packtpub.com/web-development/restful-web-api-design-nodejs-10-third-edition`)
- *Using Node.js UI Testing* (`https://www.packtpub.com/web-development/using-nodejs-ui-testing`)
- *Test-Driven Development* (`https://www.amazon.com/Test-Driven-Development-Kent-Beck/dp/0321146530/ref=sr_1_1?ie=UTF8qid=1544792837sr=8-1 keywords=tdd`)

6
Handling Requests and Responses

Now that we have created the first route, the second step is to determine what properties you will need while handling the requests that you are receiving, while also creating other routes. With that being said, it is really important to return meaningful responses so that you can change/update application states. This chapter also covers methodologies that will be helpful when testing the application, such as not directly using the request/response parameters in methods. We will also describe response status codes and when to use which status code.

Not every URL is pure and plain when requesting data. In this chapter, we are going to create a mapper that handles and calls the correct controller according to the URL's resource. Most of the time, URLs also have some query strings or additional paths for filtering data, and have a paginated output or sorting scenarios. With the help of query strings and path definitions, you will learn how to split big sources into smaller data responses, which might also be helpful in the future, in terms of web service performance.

The following topics will be covered in this chapter:

- Creating resource URIs
- Handling API responses
- URL mapping and handling query strings
- Determining whether a request was successful
- Data filtering and pagination

Technical requirements

All of the information that's required to run the code in this chapter can be found in the relevant sections of this chapter. The only requirement is that you have applications such as Node.js, VS Code, TypeScript, and so on installed on your system, which we covered in Chapter 4, *Setting Up Your Development Environment*.

All of the code that's used in this chapter is available at `https://github.com/ PacktPublishing/Hands-On-RESTful-Web-Services-with-TypeScript-3/tree/master/ Chapter06`.

Creating resource URIs

Before we get started with this chapter, we will walk through the Swagger development from `Chapter 3`, *Designing RESTful APIs with OpenAPI and Swagger*, and understand what we are going to implement.

As you may remember, the case study for Swagger is based on a simple order system, which contains two main resources called `User` and `Order`, both of which have their own operations, as shown in the following screenshot:

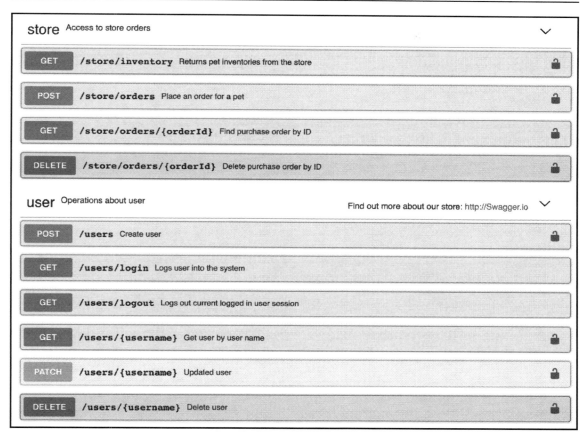

order-api resources and operations

If you don't remember the Swagger definition, go back to Chapter 3, *Designing RESTful APIs with OpenAPI and Swagger,* to go over it in detail.

The order API is a simple case study that was created to guide you throughout this book, and so there are a lot of more complex business rules when there is an order system.

The following sections will cover route and resource creation with TypeScript 3, where we will be using the base application from Chapter 5, *Building Your First API*.

Creating models

Since we know that there are two main models called User and Order, as described in the swagger file, it is time to create them for our application. To do that, go to the src folder and create a new folder, called model, like so:

```
order-api/src/model
```

In the model folder, create two files—one for each model, called user.ts and order.ts. The user.ts file's content is as follows:

```
'use strict'

export interface User {
  id: Number
  username: String
  firstName: String
  lastName: String
  email: String
  password: String
  phone: String
  userStatus: Number
}
```

Now, before we create the Order model, we have to create an enumeration to reflect the OrderStatus options. Create a file called orderStatus.ts in the same folder as the user.ts and order.ts files, with the following content:

```
'use strict'

export enum OrderStatus {
  Placed = 'PLACED',
  Approved = 'APPROVED',
  Delivered = 'DELIVERED',
}
```

Then, populate the order.ts file with the following content:

```
import { OrderStatus } from './orderStatus'

export default interface Order {
  id: Number
```

```
    userId: Number
    quantity: Number
    shipDate: Date
    status: OrderStatus
    complete: Boolean
}
```

Note that, regarding the `order.ts` file, we are importing other information from `orderStatus`, which is an enumeration, and are using it as a type on the `orderStatus` property under the `order` model.

Your project structure will look similar to the following:

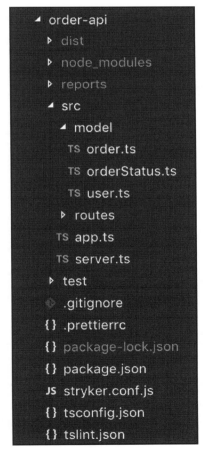

order-api structure after the models have been added

Creating tests for the missing routes

Following the same approach as in Chapter 5, *Building Your First API*, where we create some tests first, let's do the same thing here and create some tests from the missing routes, user and order.

You may remember that there is a folder called test/routes. So, in that folder, create two files, one called user.spec.ts and another called order.spec.ts.

The user.spec.ts file will contain the tests related to the user itself. Operations such as create a new user, update, retrieve, and delete will be included there. The content of the user.spec.ts file is as follows:

```
'use strict'

import * as chai from 'chai'
import chaiHttp = require('chai-http')
import 'mocha'
import app from '../../src/app'
import User from '../../src/model/User'

chai.use(chaiHttp)

const expect = chai.expect

const user: User = {
  // generic random value from 1 to 100 only for tests so far
  id: Math.floor(Math.random() * 100) + 1,
  username: 'John',
  firstName: 'John',
  lastName: 'Doe',
  email: 'jhon@myemail.com',
  password: 'password',
  phone: '5555555',
  userStatus: 1,
}

describe('userRoute', () => {
  it('should respond with HTTP 404 status because there is no user', async
() => {
    return chai
      .request(app)
      .get(`/users/${user.username}`)
      .then(res => {
        expect(res.status).to.be.equal(404)
      })
  })
```

```
it('should create a new user and retrieve it back', async () => {
  return chai
    .request(app)
    .post('/users/')
    .send(user)
    .then(res => {
      expect(res.status).to.be.equal(201)
      expect(res.body.username).to.be.equal(user.username)
    })
})
it('should return the user created on the step before', async () => {
  return chai
    .request(app)
    .get(`/users/${user.username}`)
    .then(res => {
      expect(res.status).to.be.equal(200)
      expect(res.body.username).to.be.equal(user.username)
    })
})
it('should updated the user Jhon', async () => {
  user.username = 'Jhon Updated'
  user.firstName = 'Jhon Updated'
  user.lastName = 'Doe Updated'
  user.email = 'jhon@myemail_updated.com'
  user.password = 'password Updated'
  user.phone = '3333333'
  user.userStatus = 12

  return chai
    .request(app)
    .patch(`/users/Jhon`)
    .send(user)
    .then(res => {
      expect(res.status).to.be.equal(204)
    })
})
it('should return the user updated on the step before', async () => {
  return chai
    .request(app)
    .get(`/users/${user.username}`)
    .then(res => {
      expect(res.status).to.be.equal(200)
      expect(res.body.username).to.be.equal(user.username)
      expect(res.body.firstName).to.be.equal(user.firstName)
      expect(res.body.lastName).to.be.equal(user.lastName)
      expect(res.body.email).to.be.equal(user.email)
      expect(res.body.password).to.be.equal(user.password)
      expect(res.body.phone).to.be.equal(user.phone)
```

```
      expect(res.body.userStatus).to.be.equal(user.userStatus)
    })
  })
  it('should return 404 because the user does not exist', async () => {
    user.firstName = 'Mary Jane'

    return chai
      .request(app)
      .patch(`/users/Mary`)
      .send(user)
      .then(res => {
        expect(res.status).to.be.equal(404)
      })
  })
  it('should remove an existent user', async () => {
    return chai
      .request(app)
      .del(`/users/${user.username}`)
      .then(res => {
        expect(res.status).to.be.equal(204)
      })
  })
  it('should return 404 when it is trying to remove an user because the
user does not exist', async () => {
    return chai
      .request(app)
      .del(`/users/Mary`)
      .then(res => {
        expect(res.status).to.be.equal(404)
      })
  })
})
})
```

As you can see, we created a `test` for each operation.

Imagine that the data store is empty when the test starts. This is not the best approach, but we will evolve these tests throughout this book so that we can create something that is state-of-the-art.

The scenarios that we will cover are as follows:

- Respond with a HTTP **404** status because there is no user
- Create a new user and retrieve them
- Return the user that was created in the previous step
- Update the user `John`
- Return the user that was updated in the previous step

- Return **404** because the user does not exist
- Remove an existing user
- Return **404** when you are trying to remove a user because the user does not exist

Of course, if we run these tests, they will fail because there is no implementation in place. This is exactly what we want.

We will use the same approach that we used here to order scenarios. The `order.spec.ts` file's content is as follows:

```
'use strict'

import * as chai from 'chai'
import chaiHttp = require('chai-http')
import 'mocha'
import app from '../../src/app'
import Order from '../../src/model/order'
import { OrderStatus } from '../../src/model/orderStatus'

chai.use(chaiHttp)

const expect = chai.expect

const order: Order = {
  // generic random value from 1 to 100 only for tests so far
  id: 1,
  userId: 20,
  quantity: 1,
  shipDate: new Date(),
  status: OrderStatus.Placed,
  complete: false,
}

describe('userRoute', () => {
  it('should respond with HTTP 404 status because there is no order', async
() => {
    return chai
      .request(app)
      .get(`/store/orders/${order.id}`)
      .then(res => {
        expect(res.status).to.be.equal(404)
      })
  })
  it('should create a new order and retrieve it back', async () => {
    return chai
      .request(app)
      .post('/store/orders')
```

```
        .send(order)
        .then(res => {
          expect(res.status).to.be.equal(201)
          expect(res.body.userId).to.be.equal(order.userId)
          expect(res.body.complete).to.be.equal(false)
          order.id = res.body.id
        })
    })
    it('should return the order created on the step before', async () => {
      return chai
        .request(app)
        .get(`/store/orders/${order.id}`)
        .then(res => {
          expect(res.status).to.be.equal(200)
          expect(res.body.id).to.be.equal(order.id)
          expect(res.body.status).to.be.equal(order.status)
        })
    })
    it('should return the inventory for all users', async () => {
      return chai
        .request(app)
        .get(`/store/inventory`)
        .then(res => {
          expect(res.status).to.be.equal(200)
          expect(res.body[20].length).to.be.equal(1)
        })
    })
    it('should remove an existing order', async () => {
      return chai
        .request(app)
        .del(`/store/orders/${order.id}`)
        .then(res => {
          expect(res.status).to.be.equal(204)
        })
    })
    it('should return 404 when it is trying to remove an order because the
order does not exist', async () => {
      return chai
        .request(app)
        .del(`/store/orders/${order.id}`)
        .then(res => {
          expect(res.status).to.be.equal(404)
        })
    })
  })
})
```

The same idea as before will be used for the order scenarios:

- Respond with a HTTP **404** status because there is no order
- Create a new `order` and retrieve them
- Return the `order` that was created in the previous step
- Return the inventory for all users
- Remove an existing order
- Return **404** when it is trying to remove an order because the order does not exist

Notice that, for `order`, there is a different test for inventory. Basically, we want the orders to be grouped by user, and that test will verify that.

 For the purpose of this chapter, we will consider the data to be persistent in memory.

We would also like to create a route as an index for the `order-api` application so that we can also create a test file for it, called `api.ts`, which just retrieves `title` as a GET operation:

```
import { NextFunction, Request, Response } from 'express'

export let getApi = (req: Request, res: Response, next: NextFunction) => {
  return res.status(200).send({ title: 'Order API' })
}
```

Up until now, we have tests that are failing. This is happening because there is no implementation for them. The next section is going to walk you through the code for the controllers.

Implementing controllers

Now that we have the tests in place, it is time to implement the controllers, which will take care of the operations themselves, such as adding a new user, creating a new order, removing an order, getting the inventory, and so on.

The idea is to export functions that could be used from the routes that are keeping things separated. To do that, in the `src` folder, create a folder called `controllers` and add a file called `user.ts` with the following content:

```
import { NextFunction, Request, Response } from 'express'
import { default as User } from '../model/User'

let users: Array<User> = []

export let getUser = (req: Request, res: Response, next: NextFunction) => {
  const username = req.params.username
  const user = users.find(obj => obj.username === username)
  const httpStatusCode = user ? 200 : 404
  return res.status(httpStatusCode).send(user)
}

export let addUser = (req: Request, res: Response, next: NextFunction) => {
  const user: User = {
    // generic random value from 1 to 100 only for tests so far
    id: Math.floor(Math.random() * 100) + 1,
    username: req.body.username,
    firstName: req.body.firstName,
    lastName: req.body.lastName,
    email: req.body.email,
    password: req.body.password,
    phone: req.body.phone,
    userStatus: 1,
  }
  users.push(user)
  return res.status(201).send(user)
}

export let updateUser = (req: Request, res: Response, next: NextFunction)
=> {
  const username = req.params.username
  const userIndex = users.findIndex(item => item.username === username)

  if (userIndex === -1) {
    return res.status(404).send()
  }

  const user = users[userIndex]
  user.username = req.body.username || user.username
  user.firstName = req.body.firstName || user.firstName
  user.lastName = req.body.lastName || user.lastName
  user.email = req.body.email || user.email
  user.password = req.body.password || user.password
  user.phone = req.body.phone || user.phone
```

```
  user.userStatus = req.body.userStatus || user.userStatus

  users[userIndex] = user
  return res.status(204).send()
}

export let removeUser = (req: Request, res: Response, next: NextFunction)
=> {
  const username = req.params.username
  const userIndex = users.findIndex(item => item.username === username)

  if (userIndex === -1) {
    return res.status(404).send()
  }

  users = users.filter(item => item.username !== username)

  return res.status(204).send()
}
```

Basically, we have one function for each HTTP operation. We could have more functions with a single responsibility, which is better than putting everything in a single place, but we are going to use this example to keep things simple. This example uses an in-memory list of users so that the operations manipulate this list in order to add new users and update, delete, and retrieve them.

The same approach is used in the order domain. We have a file called order.ts with the following content:

```
import { NextFunction, Request, Response } from 'express'
import * as _ from 'lodash'
import { default as Order } from '../model/order'
import { OrderStatus } from '../model/orderStatus'

let orders: Array<Order> = []

export let getOrder = (req: Request, res: Response, next: NextFunction) =>
{
  const id = req.params.id
  const order = orders.find(obj => obj.id === Number(id))
  const httpStatusCode = order ? 200 : 404
  return res.status(httpStatusCode).send(order)
}

export let addOrder = (req: Request, res: Response, next: NextFunction) =>
{
  const order: Order = {
```

```
    // generic random value from 1 to 100 only for tests so far
    id: Math.floor(Math.random() * 100) + 1,
    userId: req.body.userId,
    quantity: req.body.quantity,
    shipDate: req.body.shipDate,
    status: OrderStatus.Placed,
    complete: false,
  }
  orders.push(order)
  return res.status(201).send(order)
}

export let removeOrder = (req: Request, res: Response, next: NextFunction)
=> {
  const id = Number(req.params.id)
  const orderIndex = orders.findIndex(item => item.id === id)

  if (orderIndex === -1) {
    return res.status(404).send()
  }

  orders = orders.filter(item => item.id !== id)

  return res.status(204).send()
}

export let getInventory = (req: Request, res: Response, next: NextFunction)
=> {
  const grouppedOrders = _.groupBy(orders, 'userId')
  return res.status(200).send(grouppedOrders)
}
```

Based on the `user.ts` file, the only conceivable difference is that there is a new operation that summarizes the orders and groups them by a user who's retrieving the order as an array—something similar to the following:

```json
{
    "28": [
        {
            "id": 4,
            "userId": 28,
            "quantity": 1,
            "shipDate": "2018-12-17T01:52:19.458Z",
            "status": "PLACED",
            "complete": false
        },
        {
            "id": 99,
            "userId": 28,
            "quantity": 1,
            "shipDate": "2018-12-17T01:52:19.458Z",
            "status": "PLACED",
            "complete": false
        },
        {
            "id": 48,
            "userId": 28,
            "quantity": 1,
            "shipDate": "2018-12-17T01:52:19.458Z",
            "status": "PLACED",
            "complete": false
        }
    ],
    "30": [
        {
            "id": 79,
            "userId": 30,
            "quantity": 1,
            "shipDate": "2018-12-17T01:52:19.458Z",
            "status": "PLACED",
            "complete": false
        }
    ]
}
```

Even though we have the controllers in place, we still need to point the routes to them all. So far, your project's structure should look something like this:

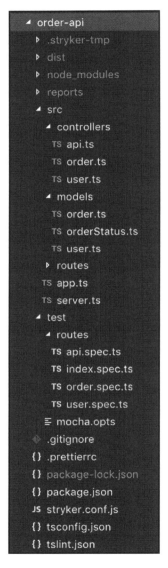

order-api structure with the tests and controllers in place

The next section is going to walk you through configuring routes and running the application that we have so far.

Configuring the remaining routes

We have the tests, configurations, models, and controllers in place, which means that we are just missing the routes for api, user, and order. To start, in the src/routes folder, create a file called api.ts with the following content:

```
import * as apiController from '../controllers/api'

export class APIRoute {
  public routes(app): void {
    app.route('/api').get(apiController.getApi)
  }
}
```

As you can see, the only difference from the default route we created in Chapter 5, *Building Your First API*, is that we are now pointing to the API controller. There are a lot of different ways to implement the routes. This method separates the controller's responsibility from the routes.

The next route we will create is user.ts:

```
import * as userController from '../controllers/user'

export class UserRoute {
  public routes(app): void {
    app.route('/users').post(userController.addUser)
    app.route('/users/:username').patch(userController.updateUser)
    app.route('/users/:username').delete(userController.removeUser)
    app.route('/users/:username').get(userController.getUser)
  }
}
```

The same idea applies here—we are just pointing each HTTP operation to its respective controller.

Finally, we will create the order.ts route:

```
import * as orderController from '../controllers/order'

export class OrderRoute {
  public routes(app): void {
    app.route('/store/inventory').get(orderController.getInventory)
    app.route('/store/orders').post(orderController.addOrder)
    app.route('/store/orders/:id').get(orderController.getOrder)
    app.route('/store/orders/:id').delete(orderController.removeOrder)
  }
}
```

The next step is to configure the `app.ts` file by adding the new routes there:

```
import * as bodyParser from 'body-parser'
import * as express from 'express'
import { APIRoute } from '../src/routes/api'
import { Index } from '../src/routes/index'
import { OrderRoute } from '../src/routes/order'
import { UserRoute } from '../src/routes/user'

class App {
  public app: express.Application
  public indexRoutes: Index = new Index()
  public userRoutes: UserRoute = new UserRoute()
  public apiRoutes: APIRoute = new APIRoute()
  public orderRoutes: OrderRoute = new OrderRoute()

  constructor() {
    this.app = express()
    this.app.use(bodyParser.json())
    this.indexRoutes.routes(this.app)
    this.userRoutes.routes(this.app)
    this.apiRoutes.routes(this.app)
    this.orderRoutes.routes(this.app)
  }
}

export default new App().app
```

Now, we are good to run our application. To do so, go to the Terminal and use the `run` command:

```
$ stryker run
```

The output will be something like the following:

File / Directory	Mutation score		# Killed	# Survived	# Timeout	# No coverage	# Runtime errors	# Transpile errors	Total detected	Total undetected	Total mutants
All files		80.56 %	80.56	77	21	10	0	20	0	87	21	128
controller/		79.31 %	79.31	60	18	9	0	18	0	69	18	105
routes/		100.00 %	100.00	16	0	1	0	0	0	17	0	17
app.ts		100.00 %	100.00	1	0	0	0	2	0	1	0	3
model/orderStatus.ts		0.00 %	0.00	0	3	0	0	0	0	0	3	3

Generated with stryker-html-reporter generator. Visit the Stryker website

Don't worry about getting a 100% **Mutation score** right now. We still need to replace the main operations so that we can use a database.

Everything looks good from a test point of view. The next section is going to test the application with Postman.

However, before we go any further, notice that there is no need to use the index route anymore. So, go into the `app.ts` file and remove it from there:

```
import * as bodyParser from 'body-parser'
import * as express from 'express'
import { APIRoute } from '../src/routes/api'
import { OrderRoute } from '../src/routes/order'
import { UserRoute } from '../src/routes/user'

class App {
  public app: express.Application
  public userRoutes: UserRoute = new UserRoute()
  public apiRoutes: APIRoute = new APIRoute()
  public orderRoutes: OrderRoute = new OrderRoute()

  constructor() {
    this.app = express()
    this.app.use(bodyParser.json())
    this.userRoutes.routes(this.app)
    this.apiRoutes.routes(this.app)
    this.orderRoutes.routes(this.app)
  }
}

export default new App().app
```

Also, delete the `src/routes/index.ts` and `test/routes/index.ts` files and run the tests again to check that everything looks good.

Running the application

Now that the application looks good, it is time to run and test it using Postman. Go to your Terminal and run the following command:

```
$ npm run dev
```

Go to the Postman application and try to create a new user by hitting the POST method with the following JSON data:

```
{
"username": "jose",
"firstName": "test",
"lastName": "test",
"email": "test",
"password": "test",
"phone": "2343432"
}
```

The following screenshot shows the result on Postman:

User being created from Postman

You should see the resource being created as the response with a HTTP status code of 201. To make sure that the resource was created, hit the GET endpoint to retrieve the user:

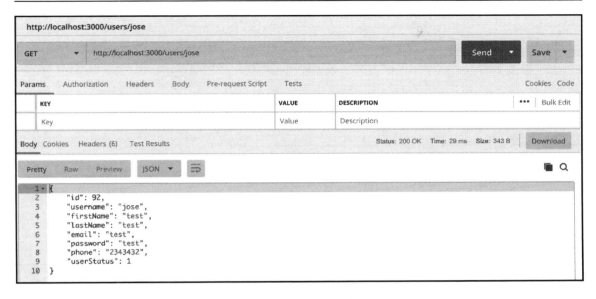

The GET endpoint getting the user jose

Now, try to create two more users. Here is our first user:

```
{
    "username": "john",
    "firstName": "john",
    "lastName": "doe",
    "email": "john@jhon.com",
    "password": "jhon",
    "phone": "5555555"
}
```

Here is our second user:

```
{
    "username": "Mary",
    "firstName": "Mary",
    "lastName": "Jane",
    "email": "mary@jane.com",
    "password": "maryjane",
    "phone": "34343434"
}
```

Now, try to GET them all. After that, update the user `john` by setting a new phone number. The user `john` is show on Postman with the following screenshot:

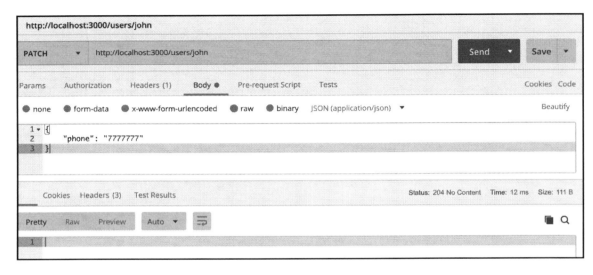

The PATCH method being used to update the user john

Right after you get the response, hit GET to see whether the phone number has changed:

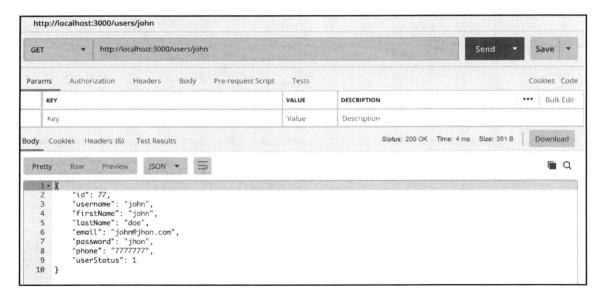

The GET method showing that the user john has changed

So far, three users have been created, and one of them was updated. To test our DELETE operation, let's try and delete the user jose by calling the DELETE operation:

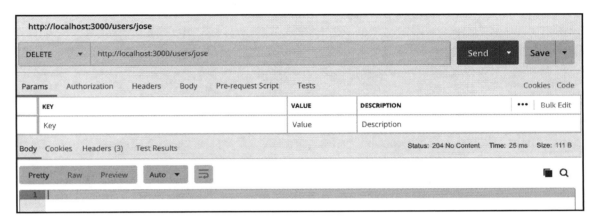

The DELETE operation with a succeed message

Now, let's try to GET the same resource:

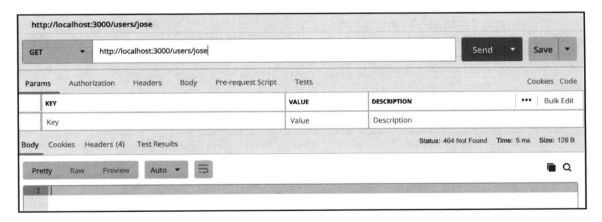

The resource jose was not found because it was removed

Everything looks good with the user route, so let's try the same approach with the order route. First, we want to create a new order with the following payload:

Note that the `userId` that you use will be different because it is a random value. So, to use it in the following requests, get the `userId` that was created by your `POST` method before and replace it in the following payload.

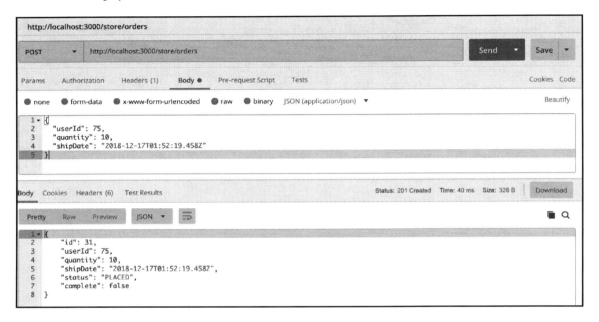

As you can see, the order was created with the status `PLACED` for the user `Mary`. Repeat the operation for the user `john`:

```
{
  "userId": 77,
  "quantity": 5,
  "shipDate": "2018-12-15T01:52:19.458Z"
}
```

Do this one more time for `Mary`:

```
{
  "userId": 75,
  "quantity": 5,
  "shipDate": "2018-12-05T01:52:19.458Z"
}
```

Now, try to get the inventory with the
endpoint's `http://localhost:3000/store/inventory` GET method. Your output will
be similar to the following:

The inventory endpoint output from Postman

If you want to see any particular order, use the `GET /store/orders/{id}` endpoint, like so:

Using the GET method to retrieve a particular order by ID

Finally, the remaining operation for `order` is the `DELETE` operation. To test it, call `DELETE`, like we did with `GET user`:

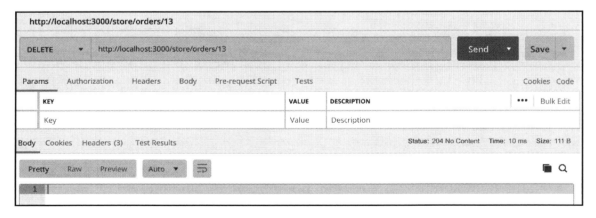

The order with an ID of 13 was removed

Just in case, hit the `GET` method again:

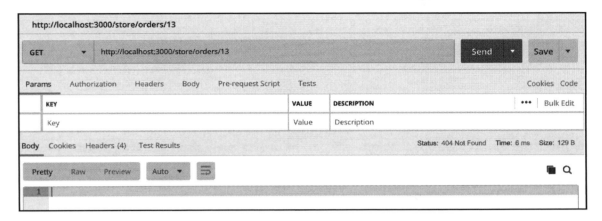

The GET method retrieving a 404 error because the resource was deleted

Just to complete all of the requests, you may want to hit the `/api` URI as well:

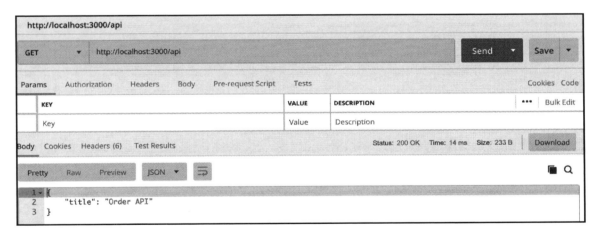

The API index URI

Query strings, data filtering, and pagination

This section is going to guide you in regards to implementing query strings, data filtering, and pagination with the resources we created in the previous sections.

Query strings and data filtering

Using query strings with TypeScript and Express.js is definitely an easy thing to do. Let's say that we would like to get the order inventory from orders that have the status of Placed. We could simply add the status query string onto the URI, for example, `http://localhost:3000/store/inventory?`**status=PLACED**, and get the server to handle everything for us.

To do that, we can just change the getInventory function in the src/controllers/order.ts file, like so:

```
export let getInventory = (req: Request, res: Response, next: NextFunction)
=> {
  const status = req.query.status
  let inventoryOrders = orders
  if (status) {
    inventoryOrders = inventoryOrders.filter(item => item.status ===
status)
  }

  const grouppedOrders = _.groupBy(inventoryOrders, 'userId')
  return res.status(200).send(grouppedOrders)
}
```

This way, every time you pass the status on the URL, you will only get the inventory for a specific order status. If there is no order in that state, the output will be empty, as follows:

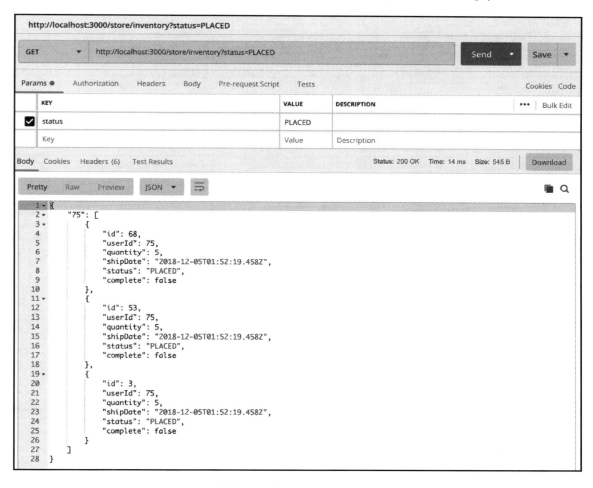

The inventory with orders in the PLACED state

The following screenshot shows the output as empty:

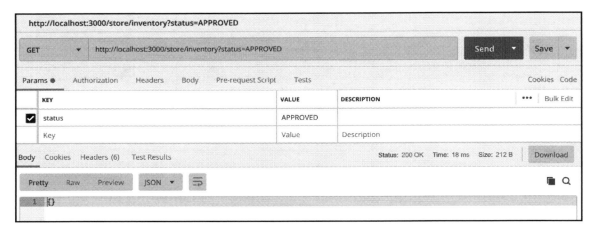

No order with the APPROVED status

As you can see, we created a simple way to filter out resources that do not match the query string.

Pagination

To create data filtering on the `order-api` service, we will expose a new endpoint that retrieves the orders through a GET method. Before we start coding, let's first change the `openapi.yml` file and add this new operation there:

```
/store/orders:
  get:
    tags:
      - store
    summary: Returns orders
    description: Returns the orders
    operationId: getOrder
    responses:
      '200':
        description: successful operation
        content:
          application/json:
            schema:
              type: object
              additionalProperties:
                type: integer
```

```
                       format: int32
        '401':
          $ref: '#/components/responses/UnauthorizedError'
      security:
        - bearerAuth: []
```

The full `openapi.yml` file is as follows:

```
openapi: 3.0.0

servers:
# Added by API Auto Mocking Plugin
  - description: SwaggerHub API Auto Mocking
    url: https://virtserver.swaggerhub.com/biharck/hands-on/1.0.0
  - description: The server description
    url: https://localhost:3000/hands-on-store/1.0.0
info:
  description: |
    This is a sample store server. You can find
    out more about Swagger at
    [http://Swagger.io](http://Swagger.io)
  version: "1.0.0"
  title: Swagger store
  termsOfService: 'http://Swagger.io/terms/'
  contact:
    email: biharck@gmail.com
  license:
    name: Apache 2.0
    url: 'http://www.apache.org/licenses/LICENSE-2.0.html'
tags:
  - name: store
    description: Access to store orders
  - name: user
    description: Operations about user
    externalDocs:
      description: Find out more about our store
      url: 'http://Swagger.io'
paths:
  /store/inventory:
    get:
      tags:
        - store
      summary: Returns user inventories from the store
      description: Returns a map of status codes to quantities
      operationId: getInventory
      responses:
        '200':
          description: successful operation
```

```
        content:
          application/json:
            schema:
              type: object
              additionalProperties:
                type: integer
                format: int32
      '401':
        $ref: '#/components/responses/UnauthorizedError'
    security:
      - bearerAuth: []
/store/orders:
  get:
    tags:
      - store
    summary: Returns orders
    description: Returns the orders
    operationId: getOrder
    responses:
      '200':
        description: successful operation
        content:
          application/json:
            schema:
              type: object
              additionalProperties:
                type: integer
                format: int32
      '401':
        $ref: '#/components/responses/UnauthorizedError'
    security:
      - bearerAuth: []
  post:
    tags:
      - store
    summary: Place an order for a user
    operationId: placeOrder
    responses:
      '201':
        description: successful operation
        content:
          application/json:
            schema:
              $ref: '#/components/schemas/Order'
          application/xml:
            schema:
              $ref: '#/components/schemas/Order'
      '400':
```

```
        description: Invalid Order
      '401':
        $ref: '#/components/responses/UnauthorizedError'
    security:
      - bearerAuth: []
    requestBody:
      content:
        application/json:
          schema:
            $ref: '#/components/schemas/Order'
      description: order placed for purchasing the user
      required: true
  '/store/orders/{orderId}':
    get:
      tags:
        - store
      summary: Find purchase order by ID
      description: >-
        For valid response try integer IDs with value >= 1 and <= 10.\ \
Other
        values will generated exceptions
      operationId: getOrderById
      parameters:
        - name: orderId
          in: path
          description: ID of user that needs to be fetched
          required: true
          schema:
            type: integer
            format: int64
            minimum: 1
            maximum: 10
      responses:
        '200':
          description: successful operation
          content:
            application/json:
              schema:
                $ref: '#/components/schemas/Order'
            application/xml:
              schema:
                $ref: '#/components/schemas/Order'
        '400':
          description: Invalid ID supplied
        '401':
          $ref: '#/components/responses/UnauthorizedError'
        '404':
          description: Order not found
```

```
      security:
        - bearerAuth: []
    delete:
      tags:
        - store
      summary: Delete purchase order by ID
      description: >-
        For valid response try integer IDs with positive integer value.\ \
        Negative or non-integer values will generate API errors
      operationId: deleteOrder
      parameters:
        - name: orderId
          in: path
          description: ID of the order that needs to be deleted
          required: true
          schema:
            type: integer
            format: int64
            minimum: 1
      responses:
        '400':
          description: Invalid ID supplied
        '401':
          $ref: '#/components/responses/UnauthorizedError'
        '404':
          description: Order not found
      security:
        - bearerAuth: []
  /users:
    post:
      tags:
        - user
      summary: Create user
      description: This can only be done by the logged in user.
      operationId: createUser
      responses:
        '201':
          description: successful operation
          content:
            application/json:
              schema:
                $ref: '#/components/schemas/User'
            application/xml:
              schema:
                $ref: '#/components/schemas/User'
        '401':
          $ref: '#/components/responses/UnauthorizedError'
      security:
```

```
      - bearerAuth: []
    requestBody:
      content:
        application/json:
          schema:
            $ref: '#/components/schemas/User'
      description: Created user object
      required: true
/users/login:
  get:
    tags:
      - user
    summary: Logs user into the system
    operationId: loginUser
    parameters:
      - name: username
        in: query
        description: The user name for login
        required: true
        schema:
          type: string
      - name: password
        in: query
        description: The password for login in clear text
        required: true
        schema:
          type: string
    responses:
      '200':
        description: successful operation
        headers:
          X-Rate-Limit:
            description: calls per hour allowed by the user
            schema:
              type: integer
              format: int32
          X-Expires-After:
            description: date in UTC when token expires
            schema:
              type: string
              format: date-time
        content:
          application/json:
            schema:
              type: string
          application/xml:
            schema:
              type: string
```

```
        '400':
          description: Invalid username/password supplied
  /users/logout:
    get:
      tags:
        - user
      summary: Logs out current logged in user session
      operationId: logoutUser
      responses:
        default:
          description: successful operation
  '/users/{username}':
    get:
      tags:
        - user
      summary: Get user by user name
      operationId: getUserByName
      parameters:
        - name: username
          in: path
          description: The name that needs to be fetched. Use user1 for
testing.
          required: true
          schema:
            type: string
      responses:
        '200':
          description: successful operation
          content:
            application/json:
              schema:
                $ref: '#/components/schemas/User'
            application/xml:
              schema:
                $ref: '#/components/schemas/User'
        '400':
          description: Invalid username supplied
        '401':
          $ref: '#/components/responses/UnauthorizedError'
        '404':
          description: User not found
      security:
        - bearerAuth: []
    patch:
      tags:
        - user
      summary: Updated user
      description: This can only be done by the logged in user.
```

```
operationId: updateUser
parameters:
  - name: username
    in: path
    description: name that need to be updated
    required: true
    schema:
      type: string
responses:
  '204':
    description: successful operation
  '400':
    description: Invalid user supplied
  '401':
    $ref: '#/components/responses/UnauthorizedError'
  '404':
    description: User not found
security:
  - bearerAuth: []
requestBody:
  content:
    application/json:
      schema:
        $ref: '#/components/schemas/User'
  description: Updated user object
  required: true
delete:
  tags:
    - user
  summary: Delete user
  description: This can only be done by the logged in user.
  operationId: deleteUser
  parameters:
    - name: username
      in: path
      description: The name that needs to be deleted
      required: true
      schema:
        type: string
  responses:
    '204':
      description: successful operation
    '400':
      description: Invalid username supplied
    '401':
      $ref: '#/components/responses/UnauthorizedError'
    '404':
      description: User not found
```

```
      security:
        - bearerAuth: []
externalDocs:
  description: Find out more about Swagger
  url: 'http://Swagger.io'
components:
  responses:
    UnauthorizedError:
      description: Access token is missing or invalid
  schemas:
    Order:
      type: object
      properties:
        id:
          type: integer
          format: int64
        userId:
          type: integer
          format: int64
        quantity:
          type: integer
          format: int32
        shipDate:
          type: string
          format: date-time
        status:
          type: string
          description: Order Status
          enum:
            - placed
            - approved
            - delivered
        complete:
          type: boolean
          default: false
      xml:
        name: Order
    User:
      type: object
      properties:
        id:
          type: integer
          format: int64
        username:
          type: string
        firstName:
          type: string
        lastName:
```

```yaml
          type: string
        email:
          type: string
        password:
          type: string
        phone:
          type: string
        userStatus:
          type: integer
          format: int32
          description: User Status
    xml:
      name: User
  securitySchemes:
    bearerAuth: # arbitrary name for the security scheme
      type: http
      scheme: bearer
      bearerFormat: JWT # optional, arbitrary value for documentation
purposes
```

Now, we can create our `test` on `test/routes/order.spec.ts`:

```typescript
it('should return all orders so far', async () => {
  return chai
    .request(app)
    .get(`/store/orders`)
    .then(res => {
      expect(res.status).to.be.equal(200)
      expect(res.body.length).to.be.equal(1)
    })
})
```

The whole file should look as follows:

```typescript
'use strict'

import * as chai from 'chai'
import chaiHttp = require('chai-http')
import 'mocha'
import app from '../../src/app'
import Order from '../../src/models/order'
import { OrderStatus } from '../../src/model/orderStatus'

chai.use(chaiHttp)

const expect = chai.expect

const order: Order = {
```

```
    // generic random value from 1 to 100 only for tests so far
    id: 1,
    userId: 20,
    quantity: 1,
    shipDate: new Date(),
    status: OrderStatus.Placed,
    complete: false,
}

describe('userRoute', () => {
  it('should respond with HTTP 404 status because there is no order', async
() => {
    return chai
      .request(app)
      .get(`/store/orders/${order.id}`)
      .then(res => {
        expect(res.status).to.be.equal(404)
      })
  })
  it('should create a new order and retrieve it back', async () => {
    return chai
      .request(app)
      .post('/store/orders')
      .send(order)
      .then(res => {
        expect(res.status).to.be.equal(201)
        expect(res.body.userId).to.be.equal(order.userId)
        expect(res.body.complete).to.be.equal(false)
        order.id = res.body.id
      })
  })
  it('should return the order created on the step before', async () => {
    return chai
      .request(app)
      .get(`/store/orders/${order.id}`)
      .then(res => {
        expect(res.status).to.be.equal(200)
        expect(res.body.id).to.be.equal(order.id)
        expect(res.body.status).to.be.equal(order.status)
      })
  })
  it('should return all orders so far', async () => {
    return chai
      .request(app)
      .get(`/store/orders`)
      .then(res => {
        expect(res.status).to.be.equal(200)
        expect(res.body.length).to.be.equal(1)
```

```
    })
  })
  it('should return the inventory for all users', async () => {
    return chai
      .request(app)
      .get(`/store/inventory`)
      .then(res => {
        expect(res.status).to.be.equal(200)
        expect(res.body[20].length).to.be.equal(1)
      })
  })
  it('should remove an existing order', async () => {
    return chai
      .request(app)
      .del(`/store/orders/${order.id}`)
      .then(res => {
        expect(res.status).to.be.equal(204)
      })
  })
  it('should return 404 when it is trying to remove an order because the
order does not exist', async () => {
    return chai
      .request(app)
      .del(`/store/orders/${order.id}`)
      .then(res => {
        expect(res.status).to.be.equal(404)
      })
  })
  })
})
```

Now, we can implement the GET method, including the pagination rule on
the src/controllers/order.ts file:

```
import { NextFunction, Request, Response } from 'express'
import * as _ from 'lodash'
import { default as Order } from '../models/order'
import { OrderStatus } from '../models/orderStatus'

let orders: Array<Order> = []

export let getOrder = (req: Request, res: Response, next: NextFunction) =>
{
  const id = req.params.id
  const order = orders.find(obj => obj.id === Number(id))
  const httpStatusCode = order ? 200 : 404
  return res.status(httpStatusCode).send(order)
}
```

```
export let getAllOrders = (req: Request, res: Response, next: NextFunction)
=> {
  return res.status(200).send(orders)
}

export let addOrder = (req: Request, res: Response, next: NextFunction) =>
{
  const order: Order = {
    // generic random value from 1 to 100 only for tests so far
    id: Math.floor(Math.random() * 100) + 1,
    userId: req.body.userId,
    quantity: req.body.quantity,
    shipDate: req.body.shipDate,
    status: OrderStatus.Placed,
    complete: false,
  }
  orders.push(order)
  return res.status(201).send(order)
}

export let removeOrder = (req: Request, res: Response, next: NextFunction)
=> {
  const id = Number(req.params.id)
  const orderIndex = orders.findIndex(item => item.id === id)

  if (orderIndex === -1) {
    return res.status(404).send()
  }

  orders = orders.filter(item => item.id !== id)

  return res.status(204).send()
}

export let getInventory = (req: Request, res: Response, next: NextFunction)
=> {
  const status = req.query.status
  let inventoryOrders = orders
  if (status) {
    inventoryOrders = inventoryOrders.filter(item => item.status ===
status)
  }

  const grouppedOrders = _.groupBy(inventoryOrders, 'userId')
  return res.status(200).send(grouppedOrders)
}
```

We can also include the route on `src/routes/order.ts`:

```
import * as orderController from '../controllers/order'

export class OrderRoute {
  public routes(app): void {
    app.route('/store/inventory').get(orderController.getInventory)
    app.route('/store/orders').post(orderController.addOrder)
    app.route('/store/orders').get(orderController.getAllOrders)
    app.route('/store/orders/:id').get(orderController.getOrder)
    app.route('/store/orders/:id').delete(orderController.removeOrder)
  }
}
```

Now, if you start the application and add some order and then hit the new GET method, you should be able to see all of the orders:

Example of new GET operation in action

Now that this has been implemented, the next step is to add a new test that will use the `limit` and `offset` strategies.

On `test/routes/order.ts`, add a new test:

```
it('should not return orders because offset is higher than the size of the
orders array', async () => {
    return chai
      .request(app)
      .get(`/store/orders?offset=2&limit=2`)
      .then(res => {
        expect(res.status).to.be.equal(200)
        expect(res.body.length).to.be.equal(0)
      })
  })
```

And in the `src/controllers/orders.ts` file, update the `getAllOrders` function with the new query strings:

```
it('should not return orders because offset is higher than the size of the
orders array', async () => {
    return chai
      .request(app)
      .get(`/store/orders?offset=2&limit=2`)
      .then(res => {
        expect(res.status).to.be.equal(200)
        expect(res.body.length).to.be.equal(0)
      })
  })
```

For that, we are using `loadash` again, which makes our lives easier.

After running the tests, you should see that they all pass:

```
$ npm run test
```

The output should be as follows:

```
baseRoute
    should respond with HTTP 200 status (48ms)
    should respond with success message

  userRoute
    should respond with HTTP 404 status because there is no order
    should create a new order and retrieve it back
    should return the order created on the step before
    should return all orders so far
    should not return orders because offset is higher than the size of the
```

```
orders array
    should return the inventory for all users
    should remove an existing order
    should return 404 when it is trying to remove an order because the
order does not exist

  userRoute
    should respond with HTTP 404 status because there is no user
    should create a new user and retrieve it back
    should return the user created on the step before
    should updated the user John
    should return the user updated on the step before
    should return 404 because the user does not exist
    should remove an existent user
    should return 404 when it is trying to remove an user because the user
does not exist


  18 passing (138ms)
```

Now, if you start the application, create a few orders, and use the pagination strategy, everything should work fine:

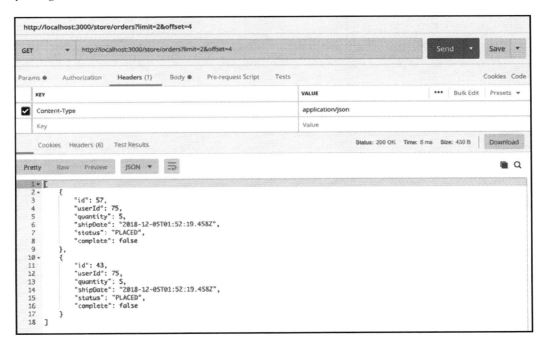

An example with limit and offset when getting orders

Summary

This chapter was an interesting one, in which you put your coding skills to practice. You were able to learn how to create the routes we defined in Swagger in Chapter 3, *Designing RESTful APIs with OpenAPI and Swagger,* so that the application matches the specification.

You also learned how to code starting, with tests and how to validate them with mutation tests by using Stryker.

In addition to this you were able to create a new operation to retrieve orders and apply filter and pagination strategies to that.

The next chapter, Chapter 7, *Formatting the API – Output,* will walk you through how to format the service output, because having a standard data format for responses is one of the key points of web service stability.

Questions

1. When is the best time to write tests?
2. What does it mean to model?
3. How do you get a query string from requests?
4. How many query parameters can you use from requests?
5. What should a route file look like?
6. Why should you separate the routes from the controller?
7. What should the default behavior from the API to respond if the offset is higher than the number of elements?

Further reading

To improve your knowledge of what was covered in this chapter, the following books are recommended, as they will be helpful for the upcoming chapters:

- *RESTful Web API Design with Node.js* (https://www.packtpub.com/web-development/restful-web-api-design-nodejs-10-third-edition)
- *Using Node.js UI Testing* (https://www.packtpub.com/web-development/using-nodejs-ui-testing)
- *Test-Driven Development* (https://www.amazon.com/Test-Driven-Development-Kent-Beck/dp/0321146530/ref=sr_1_1?ie=UTF8qid=1544792837sr=8-1 keywords=tdd)

Formatting the API - Output 7

Having a standard data format for responses is one of the key points for ensuring web service stability. In this chapter, we will introduce content negotiation, output formats, and the HAL JSON format to explain stateless API conventions. In the *Data serialization* section, you will learn how to convert resource objects into JSON objects or JSON arrays. Since JSON has become a standard, we will mainly focus on that format. We will also talk about how to expose data as XML files.

The following topics will be covered in this chapter:

- Content negotiation
- Data serialization
- Other output types

Technical requirements

All of the information that's required to run the code in this chapter can be found in the relevant sections of this chapter. The only requirement is that you have applications such as Node.js, VS Code and TypeScript installed on your system, which we covered in `Chapter 4`, *Setting Up Your Development Environment*.

All of the code that's used in this chapter is available at `https://github.com/ PacktPublishing/Hands-On-RESTful-Web-Services-with-TypeScript-3/tree/master/ Chapter07`.

Content negotiation

The main definition of content negotiation is that it's an HTTP mechanism that enables different versions of specific representations of a resource at the same URI. Using this approach, the requester is able to specify which version fits their requirements. Of course, with TypeScript, it couldn't be different and you can easily enable that possibility.

One common way to specify the response body format is passing the payload at the header of the HTTP request, like so:

```
Accept: application/javascript
```

The following screenshot shows an example from Postman with the accept header as `application/javascript`:

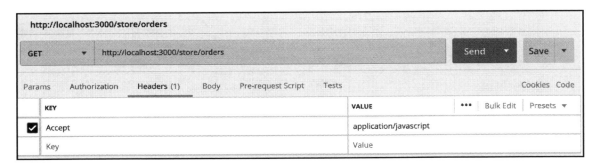

Accept header definition using Postman for application/JavaScript format

Based on the previous screenshot, the server is not prepared to retrieve the data for the GET operation, that is, `application/javascript`. Hence, it should return a **406** error, as shown in the following screenshot:

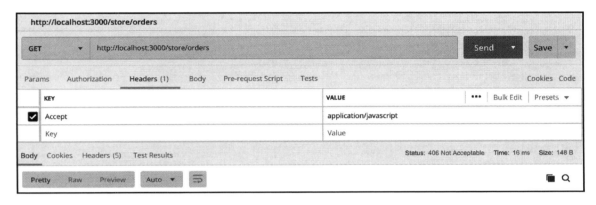

Example of Accept not being allowed

A simple way to implement this on the server side is by defining the response format. Let's look at an example using the `getAllOrders` operation from `src/controllers/orders.ts`:

```
export let getAllOrders = (req: Request, res: Response, next: NextFunction)
=> {
  const limit = req.query.limit || orders.length
  const offset = req.query.offset || 0
  return res.status(200).send(
    _(orders)
      .drop(offset)
      .take(limit)
      .value()
  )
}
```

We want to specify that this operation comes from the GET and that orders will only be returned in JSON format. To do so, let's add a response format:

```
...
const APPLICATION_JSON = 'application/json'
...
export let getAllOrders = (req: Request, res: Response, next: NextFunction)
=> {
  const limit = req.query.limit || orders.length
  const offset = req.query.offset || 0
```

```
const filteredOrders = _(orders)
  .drop(offset)
  .take(limit)
  .value()

res.format({
  json: () => {
    res.type(APPLICATION_JSON)
    res.status(200).send(filteredOrders)
  }
})
}
```

Now, we have a response format in place for JSON. The next step is to add a default behavior when `Accept` comes with a different format, such as `application/xml`:

```
export let getAllOrders = (req: Request, res: Response, next: NextFunction)
=> {
  const limit = req.query.limit || orders.length
  const offset = req.query.offset || 0

  const filteredOrders = _(orders)
    .drop(offset)
    .take(limit)
    .value()

  res.format({
    json: () => {
      res.type(APPLICATION_JSON)
      res.status(200).send(filteredOrders)
    },
    default: () => {
      res.status(406).send()
    }
  })
}
```

Now, restart the server running `npm run dev` the call `GET` with `application/xml`, we should see the **406 Not Acceptable**, HTTP status:

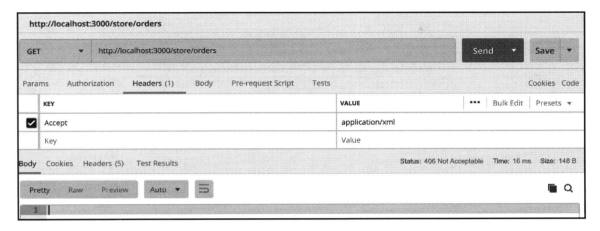

Server returning 406 because XML is not allowed

Now, we can go even further and apply this rule to all of the operations we have so far for order, api, and user.

The following code block shows the changes for the api controller:

```
import { NextFunction, Request, Response } from 'express'
const APPLICATION_JSON = 'application/json'

export let getApi = (req: Request, res: Response, next: NextFunction) => {
  return res.format({
    json: () => {
      res.type(APPLICATION_JSON)
      res.status(200).send({ title: 'Order API' })
    },
    default: () => {
      res.status(406).send()
    },
  })
}
```

The following code block shows the changes for the user controller:

```
import { NextFunction, Request, Response } from 'express'
import { default as User } from '../models/User'
const APPLICATION_JSON = 'application/json'

let users: Array<User> = []

export let getUser = (req: Request, res: Response, next: NextFunction) => {
  const username = req.params.username
```

```
  const user = users.find(obj => obj.username === username)
  const httpStatusCode = user ? 200 : 404
  return res.format({
    json: () => {
      res.type(APPLICATION_JSON)
      res.status(httpStatusCode).send(user)
    },
    default: () => {
      res.status(406).send()
    },
  })
}

export let addUser = (req: Request, res: Response, next: NextFunction) => {
  const user: User = {
    // generic random value from 1 to 100 only for tests so far
    id: Math.floor(Math.random() * 100) + 1,
    username: req.body.username,
    firstName: req.body.firstName,
    lastName: req.body.lastName,
    email: req.body.email,
    password: req.body.password,
    phone: req.body.phone,
    userStatus: 1,
  }
  users.push(user)
  return res.format({
    json: () => {
      res.type(APPLICATION_JSON)
      res.status(201).send(user)
    },
    default: () => {
      res.status(406).send()
    },
  })
}

export let updateUser = (req: Request, res: Response, next: NextFunction)
=> {
  const username = req.params.username
  const userIndex = users.findIndex(item => item.username === username)

  if (userIndex === -1) {
    return res.status(404).send()
  }

  const user = users[userIndex]
  user.username = req.body.username || user.username
```

```
      user.firstName = req.body.firstName || user.firstName
      user.lastName = req.body.lastName || user.lastName
      user.email = req.body.email || user.email
      user.password = req.body.password || user.password
      user.phone = req.body.phone || user.phone
      user.userStatus = req.body.userStatus || user.userStatus

      users[userIndex] = user
      return res.format({
        json: () => {
          res.type(APPLICATION_JSON)
          res.status(204).send()
        },
        default: () => {
          res.status(406).send()
        },
      })
}

export let removeUser = (req: Request, res: Response, next: NextFunction)
=> {
    const username = req.params.username
    const userIndex = users.findIndex(item => item.username === username)

    if (userIndex === -1) {
      return res.status(404).send()
    }

    users = users.filter(item => item.username !== username)

    return res.format({
      json: () => {
        res.type(APPLICATION_JSON)
        res.status(204).send()
      },
      default: () => {
        res.status(406).send()
      },
    })
}
```

The following code block shows the changes for the `order` controller:

```
import { NextFunction, Request, Response } from 'express'
import * as _ from 'lodash'
import { default as Order } from '../models/order'
import { OrderStatus } from '../models/orderStatus'
const APPLICATION_JSON = 'application/json'
```

```
let orders: Array<Order> = []

export let getOrder = (req: Request, res: Response, next: NextFunction) =>
{
  const id = req.params.id
  const order = orders.find(obj => obj.id === Number(id))
  const httpStatusCode = order ? 200 : 404
  return res.status(httpStatusCode).send(order)
}

export let getAllOrders = (req: Request, res: Response, next: NextFunction)
=> {
  const limit = req.query.limit || orders.length
  const offset = req.query.offset || 0

  const filteredOrders = _(orders)
    .drop(offset)
    .take(limit)
    .value()

  res.format({
    json: () => {
      res.type(APPLICATION_JSON)
      res.status(200).send(filteredOrders)
    },
    default: () => {
      res.status(406).send()
    },
  })
}

export let addOrder = (req: Request, res: Response, next: NextFunction) =>
{
  const order: Order = {
    // generic random value from 1 to 100 only for tests so far
    id: Math.floor(Math.random() * 100) + 1,
    userId: req.body.userId,
    quantity: req.body.quantity,
    shipDate: req.body.shipDate,
    status: OrderStatus.Placed,
    complete: false,
  }

  orders.push(order)

  return res.format({
    json: () => {
      res.type(APPLICATION_JSON)
```

```
      res.status(201).send(order)
    },
    default: () => {
      res.status(406).send()
    },
  })
}

export let removeOrder = (req: Request, res: Response, next: NextFunction)
=> {
  const id = Number(req.params.id)
  const orderIndex = orders.findIndex(item => item.id === id)

  if (orderIndex === -1) {
    return res.status(404).send()
  }

  orders = orders.filter(item => item.id !== id)

  return res.format({
    json: () => {
      res.type(APPLICATION_JSON)
      res.status(204).send()
    },
    default: () => {
      res.status(406).send()
    },
  })
}

export let getInventory = (req: Request, res: Response, next: NextFunction)
=> {
  const status = req.query.status
  let inventoryOrders = orders
  if (status) {
    inventoryOrders = inventoryOrders.filter(item => item.status ===
status)
  }

  const grouppedOrders = _.groupBy(inventoryOrders, 'userId')

  return res.format({
    json: () => {
      res.type(APPLICATION_JSON)
      res.status(200).send(grouppedOrders)
    },
    default: () => {
      res.status(406).send()
```

```
    },
  })
}
```

As you can see, there's a lot of duplicate code, so let's move it to a utility class called `orderApiUtility.ts`, **under** `src/utility`:

```
import { Response } from 'express'
import { ApplicationType } from '../models/applicationType'

export let formatOutput = (
  res: Response,
  data: any,
  statusCode: number,
  applicationType: ApplicationType
) => {
  return res.format({
    json: () => {
      res.type(applicationType)
      res.status(statusCode).send(data)
    },
    default: () => {
      res.status(406).send()
    },
  })
}
```

Notice that we are using an `enum` called `ApplicationType`. Go into the `models` folder and create a file called `applicationType.ts` with the following content:

```
'use strict'

export enum ApplicationType {
  JSON = 'application/json',
  XML = 'application/xml',
}
```

 Up until now, we only cared about using JSON and XML, but feel free to add any other media type you want to.

Now, the `api` controller should be as follows:

```
import { NextFunction, Request, Response } from 'express'
import { ApplicationType } from '../models/applicationType'
import { formatOutput } from '../utility/orderApiUtility'
```

```
export let getApi = (req: Request, res: Response, next: NextFunction) => {
  return formatOutput(res, { title: 'Order API' }, 200,
ApplicationType.JSON)
}
```

The following code block shows the `user` controller:

```
import { NextFunction, Request, Response } from 'express'
import { ApplicationType } from '../models/applicationType'
import { default as User } from '../models/User'
import { formatOutput } from '../utility/orderApiUtility'
const APPLICATION_JSON = 'application/json'

let users: Array<User> = []

export let getUser = (req: Request, res: Response, next: NextFunction) => {
  const username = req.params.username
  const user = users.find(obj => obj.username === username)
  const httpStatusCode = user ? 200 : 404

  return formatOutput(res, user, httpStatusCode, ApplicationType.JSON)
}

export let addUser = (req: Request, res: Response, next: NextFunction) => {
  const user: User = {
    // generic random value from 1 to 100 only for tests so far
    id: Math.floor(Math.random() * 100) + 1,
    username: req.body.username,
    firstName: req.body.firstName,
    lastName: req.body.lastName,
    email: req.body.email,
    password: req.body.password,
    phone: req.body.phone,
    userStatus: 1,
  }
  users.push(user)
  return formatOutput(res, user, 201, ApplicationType.JSON)
}

export let updateUser = (req: Request, res: Response, next: NextFunction)
=> {
  const username = req.params.username
  const userIndex = users.findIndex(item => item.username === username)

  if (userIndex === -1) {
    return res.status(404).send()
  }
```

```
    const user = users[userIndex]
    user.username = req.body.username || user.username
    user.firstName = req.body.firstName || user.firstName
    user.lastName = req.body.lastName || user.lastName
    user.email = req.body.email || user.email
    user.password = req.body.password || user.password
    user.phone = req.body.phone || user.phone
    user.userStatus = req.body.userStatus || user.userStatus

    users[userIndex] = user
    return formatOutput(res, {}, 204, ApplicationType.JSON)
}

export let removeUser = (req: Request, res: Response, next: NextFunction)
=> {
  const username = req.params.username
  const userIndex = users.findIndex(item => item.username === username)

  if (userIndex === -1) {
    return res.status(404).send()
  }

  users = users.filter(item => item.username !== username)
  return formatOutput(res, {}, 204, ApplicationType.JSON)
}
```

The following code block shows the `order` controller:

```
import { NextFunction, Request, Response } from 'express'
import * as _ from 'lodash'
import { default as Order } from '../models/order'
import { OrderStatus } from '../models/orderStatus'
const APPLICATION_JSON = 'application/json'
import { ApplicationType } from '../models/applicationType'
import { formatOutput } from '../utility/orderApiUtility'

let orders: Array<Order> = []

export let getOrder = (req: Request, res: Response, next: NextFunction) =>
{
  const id = req.params.id
  const order = orders.find(obj => obj.id === Number(id))
  const httpStatusCode = order ? 200 : 404
  return formatOutput(res, order, httpStatusCode, ApplicationType.JSON)
}

export let getAllOrders = (req: Request, res: Response, next: NextFunction)
=> {
```

```
    const limit = req.query.limit || orders.length
    const offset = req.query.offset || 0

    const filteredOrders = _(orders)
      .drop(offset)
      .take(limit)
      .value()

    return formatOutput(res, filteredOrders, 200, ApplicationType.JSON)

}

export let addOrder = (req: Request, res: Response, next: NextFunction) =>
{
    const order: Order = {
      // generic random value from 1 to 100 only for tests so far
      id: Math.floor(Math.random() * 100) + 1,
      userId: req.body.userId,
      quantity: req.body.quantity,
      shipDate: req.body.shipDate,
      status: OrderStatus.Placed,
      complete: false,
    }

    orders.push(order)

    return formatOutput(res, order, 201, ApplicationType.JSON)

}

export let removeOrder = (req: Request, res: Response, next: NextFunction)
=> {
    const id = Number(req.params.id)
    const orderIndex = orders.findIndex(item => item.id === id)

    if (orderIndex === -1) {
      return res.status(404).send()
    }

    orders = orders.filter(item => item.id !== id)

    return formatOutput(res, {}, 204, ApplicationType.JSON)

}

export let getInventory = (req: Request, res: Response, next: NextFunction)
=> {
    const status = req.query.status
```

```
    let inventoryOrders = orders
    if (status) {
      inventoryOrders = inventoryOrders.filter(item => item.status ===
  status)
    }

    const grouppedOrders = _.groupBy(inventoryOrders, 'userId')

    return formatOutput(res, grouppedOrders, 200, ApplicationType.JSON)

  }
```

Let's run the tests again and see the output:

```
$ npm run test
```

The output should be something like the following:

```
baseRoute
    should respond with HTTP 200 status (45ms)
    should respond with success message

  userRoute
    should respond with HTTP 404 status because there is no order
    should create a new order and retrieve it back
    should return the order created on the step before
    should return all orders so far
    should not return orders because offset is higher than the size of the
orders array
    should return the inventory for all users
    should remove an existing order
    should return 404 when it is trying to remove an order because the
order does not exist

  userRoute
    should respond with HTTP 404 status because there is no user
    should create a new user and retrieve it back
    should return the user created on the step before
    should updated the user John
    should return the user updated on the step before
    should return 404 because the user does not exist
    should remove an existent user
    should return 404 when it is trying to remove an user because the user
does not exist


    18 passing (166ms)
```

This shows us that everything looks good and we are good to move onto the next section, which is all about data serialization.

Data serialization

If you walk through the `npm` repository, you might see a lot of packages that do data serialization. This section uses a library called `js2xmlparser` to show you various possibilities, such as providing a way to encode data—not only as JSON, but also as XML.

 Remember that you can decide what flavor of framework you want to use.

To keep enriching our `order-api` application, imagine that you have a new request to provide support for XML encoding of request and response bodies as well. You may not want to duplicate the routes and controller; first, because it is not the best solution, and second, because you don't want to. One of the easiest ways to implement it is by using external packages such as `js2xmlparser`, as we mentioned previously.

To use `js2xmlparser`, install the package for it in the `order-api` application with the following command:

```
$ npm install js2xmlparser --save
```

 `js2xmlparser` is available at `https://www.npmjs.com/package/js2xmlparser`.

After that, your `package.json` file should look as follows:

```
{
  "name": "order-api",
  "version": "1.0.0",
  "description": "This is the example from the Book Hands on RESTful Web
Services with TypeScript 3",
  "main": "./dist/server.js",
  "scripts": {
    "build": "npm run clean && tsc",
    "clean": "rimraf dist && rimraf reports",
    "lint": "tslint ./src/**/*.ts ./test/**/*.spec.ts",
    "lint:fix": "tslint --fix ./src/**/*.ts ./test/**/*.spec.ts -t
```

```
verbose",
    "pretest": "cross-env NODE_ENV=test npm run build && npm run lint",
    "test": "cross-env NODE_ENV=test mocha --reporter spec --compilers
ts:ts-node/register test/**/*.spec.ts ",
    "test:mutation": "stryker run",
    "stryker:init": "stryker init",
    "dev": "cross-env PORT=3000 NODE_ENV=dev ts-node ./src/server.ts",
    "prod": "PORT=3000 npm run build && npm run start",
    "tsc": "tsc"
  },
  "engines": {
    "node": ">=8.0.0"
  },
  "keywords": [
    "order POC",
    "Hands on RESTful Web Services with TypeScript 3",
    "TypeScript 3",
    "Packt Books"
  ],
  "author": "Biharck Muniz Araújo",
  "license": "MIT",
  "devDependencies": {
    "@types/body-parser": "^1.17.0",
    "@types/lodash",
    "chai-http",
    "@types/chai": "^4.1.7",
    "@types/chai-http": "^3.0.5",
    "@types/express": "^4.16.0",
    "@types/mocha": "^5.2.5",
    "@types/node": "^10.12.12",
    "chai": "^4.2.0",
    "cross-env": "^5.2.0",
    "mocha": "^5.2.0",
    "rimraf": "^2.6.2",
    "stryker": "^0.33.1",
    "stryker-api": "^0.22.0",
    "stryker-html-reporter": "^0.16.9",
    "stryker-mocha-framework": "^0.13.2",
    "stryker-mocha-runner": "^0.15.2",
    "stryker-typescript": "^0.16.1",
    "ts-node": "^7.0.1",
    "tslint": "^5.11.0",
    "tslint-config-prettier": "^1.17.0",
    "typescript": "^3.2.1"
  },
  "dependencies": {
    "body-parser": "^1.18.3",
    "express": "^4.16.4",
```

```
  "js2xmlparser": "^3.0.0",
  "lodash": "^4.17.11"
  }
}
```

Now, we are good to implement the changes. Let's start with the src/util/orderApiUtility.ts file:

```typescript
import { Response } from 'express'
import * as js2xmlparser from 'js2xmlparser'
import { ApplicationType } from '../models/applicationType'

export let formatOutput = (
  res: Response,
  data: any,
  statusCode: number,
  rootElement?: string
) => {
  return res.format({
    json: () => {
      res.type(ApplicationType.JSON)
      res.status(statusCode).send(data)
    },
    xml: () => {
      res.type(ApplicationType.XML)
      res.status(200).send(js2xmlparser.parse(rootElement, data))
    },
    default: () => {
      res.status(406).send()
    },
  })
}
```

As you can see, we included the js2xmlparser package and removed the Application type as a parameter, since all operations will return either JSON or XML due to their acceptance criteria. We also added an optional parameter, called rootElement, which is responsible for describing the root XML element's name.

The next step is to change the controllers (api, user, and order). We'll start with the api controller:

```typescript
import { NextFunction, Request, Response } from 'express'
import { formatOutput } from '../utility/orderApiUtility'

export let getApi = (req: Request, res: Response, next: NextFunction) => {
  return formatOutput(res, { title: 'Order API' }, 200)
}
```

Next, we'll move on to the `user` controller:

```
import { NextFunction, Request, Response } from 'express'
import { default as User } from '../models/user'
import { formatOutput } from '../utility/orderApiUtility'

let users: Array<User> = []

export let getUser = (req: Request, res: Response, next: NextFunction) => {
  const username = req.params.username
  const user = users.find(obj => obj.username === username)
  const httpStatusCode = user ? 200 : 404

  return formatOutput(res, user, httpStatusCode, 'user')
}

export let addUser = (req: Request, res: Response, next: NextFunction) => {
  const user: User = {
    // generic random value from 1 to 100 only for tests so far
    id: Math.floor(Math.random() * 100) + 1,
    username: req.body.username,
    firstName: req.body.firstName,
    lastName: req.body.lastName,
    email: req.body.email,
    password: req.body.password,
    phone: req.body.phone,
    userStatus: 1,
  }
  users.push(user)
  return formatOutput(res, user, 201, 'user')
}

export let updateUser = (req: Request, res: Response, next: NextFunction)
=> {
  const username = req.params.username
  const userIndex = users.findIndex(item => item.username === username)

  if (userIndex === -1) {
    return res.status(404).send()
  }

  const user = users[userIndex]
  user.username = req.body.username || user.username
  user.firstName = req.body.firstName || user.firstName
  user.lastName = req.body.lastName || user.lastName
  user.email = req.body.email || user.email
  user.password = req.body.password || user.password
  user.phone = req.body.phone || user.phone
```

```
  user.userStatus = req.body.userStatus || user.userStatus

  users[userIndex] = user
  return formatOutput(res, {}, 204)
}

export let removeUser = (req: Request, res: Response, next: NextFunction)
=> {
  const username = req.params.username
  const userIndex = users.findIndex(item => item.username === username)

  if (userIndex === -1) {
    return res.status(404).send()
  }

  users = users.filter(item => item.username !== username)
  return formatOutput(res, {}, 204)
}
```

We'll then move on to the order controller:

```
import { NextFunction, Request, Response } from 'express'
import * as _ from 'lodash'
import { default as Order } from '../models/order'
import { OrderStatus } from '../models/orderStatus'
import { formatOutput } from '../utility/orderApiUtility'

let orders: Array<Order> = []

export let getOrder = (req: Request, res: Response, next: NextFunction) =>
{
  const id = req.params.id
  const order = orders.find(obj => obj.id === Number(id))
  const httpStatusCode = order ? 200 : 404
  return formatOutput(res, order, httpStatusCode, 'order')
}

export let getAllOrders = (req: Request, res: Response, next: NextFunction)
=> {
  const limit = req.query.limit || orders.length
  const offset = req.query.offset || 0

  const filteredOrders = _(orders)
    .drop(offset)
    .take(limit)
    .value()

  return formatOutput(res, filteredOrders, 200, 'order')
```

```
}

export let addOrder = (req: Request, res: Response, next: NextFunction) =>
{
  const order: Order = {
    // generic random value from 1 to 100 only for tests so far
    id: Math.floor(Math.random() * 100) + 1,
    userId: req.body.userId,
    quantity: req.body.quantity,
    shipDate: req.body.shipDate,
    status: OrderStatus.Placed,
    complete: false,
  }

  orders.push(order)

  return formatOutput(res, order, 201, 'order')
}

export let removeOrder = (req: Request, res: Response, next: NextFunction)
=> {
  const id = Number(req.params.id)
  const orderIndex = orders.findIndex(item => item.id === id)

  if (orderIndex === -1) {
    return res.status(404).send()
  }

  orders = orders.filter(item => item.id !== id)

  return formatOutput(res, {}, 204)
}

export let getInventory = (req: Request, res: Response, next: NextFunction)
=> {
  const status = req.query.status
  let inventoryOrders = orders
  if (status) {
    inventoryOrders = inventoryOrders.filter(item => item.status ===
status)
  }

  const grouppedOrders = _.groupBy(inventoryOrders, 'userId')

  return formatOutput(res, grouppedOrders, 200, 'inventory')
}
```

Now, if you run the GET user for the instance that has an acceptance of application/xml, you should see the following content, (the Accept header should be set):

```xml
<?xml version='1.0'?>
<user>
    <id>85</id>
    <username>Mary</username>
    <firstName>Mary</firstName>
    <lastName>Jane</lastName>
    <email>mary@jane.com</email>
    <password>maryjane</password>
    <phone>34343434</phone>
    <userStatus>1</userStatus>
</user>
```

The preceding code snippet shows a GET request as XML in the following screenshot:

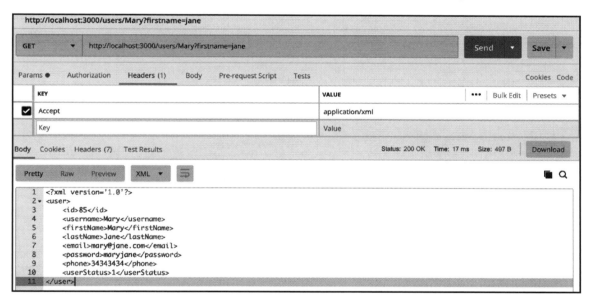

GET user URI in XML

The same idea applies to `application/json`:

```
{
    "id": 85,
    "username": "Mary",
    "firstName": "Mary",
    "lastName": "Jane",
    "email": "mary@jane.com",
    "password": "maryjane",
    "phone": "34343434",
    "userStatus": 1
}
```

The result of the preceding code snippet can be seen in the following screenshot:

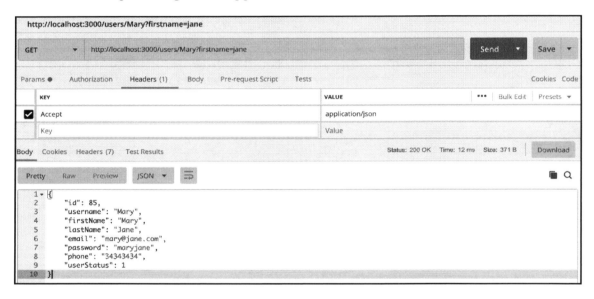

GET user URI as JSON

However, if you try any other type of acceptance, you should see an HTTP status of **406 Not Acceptable**, as follows:

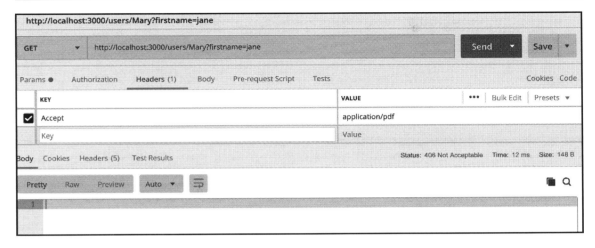

Request not acceptable for application/pdf format

Hypertext Application Language

As already mentioned in this Book, **Hypertext Application Language (HAL)** is a standard convention for defining hypermedia such as links to external resources. The idea of this section is to show you how to enable this feature.

In order to implement HAL in our `order-api` application, we will need to install a package called `halson`.

Just like with the other packages mentioned in this book, you can choose anyone you like.

`halson` is available at `https://www.npmjs.com/package/halson`.

The installation is simple, and it's just like any other `npm` package:

```
$ npm install --save halson
```

Now that you've installed `halson`, change the `getOrder` method, like so:

```
export let getOrder = (req: Request, res: Response, next: NextFunction) =>
{
  const id = req.params.id
  let order = orders.find(obj => obj.id === Number(id))
  const httpStatusCode = order ? 200 : 404
  order = halson(order).addLink('self', `/store/orders/${order.id}`)
  return formatOutput(res, order, httpStatusCode, 'order')
}
```

The change demonstrates how we create a resource with `halson`. It is very simple— First we create a HALSON resource by passing the resource data to the halson() function. After that, we add a link named "self" using the addLink() method, which returns the modified HALSON resource. The second argument of the addLink() method is a relative or absolute URL specifying where the link should point to. The full `order.ts` file is as follows:

```
import { NextFunction, Request, Response } from 'express'
import * as halson from 'halson'
import * as _ from 'lodash'
import { default as Order } from '../models/order'
import { OrderStatus } from '../models/orderStatus'
import { formatOutput } from '../utility/orderApiUtility'

let orders: Array<Order> = []

export let getOrder = (req: Request, res: Response, next: NextFunction) =>
{
  const id = req.params.id
  let order = orders.find(obj => obj.id === Number(id))
  const httpStatusCode = order ? 200 : 404
  order = halson(order).addLink('self', `/store/orders/${order.id}`)
  return formatOutput(res, order, httpStatusCode, 'order')
}

export let getAllOrders = (req: Request, res: Response, next: NextFunction)
=> {
  const limit = req.query.limit || orders.length
  const offset = req.query.offset || 0

  const filteredOrders = _(orders)
    .drop(offset)
    .take(limit)
    .value()

  return formatOutput(res, filteredOrders, 200, 'order')
}
```

```
export let addOrder = (req: Request, res: Response, next: NextFunction) =>
{
  const order: Order = {
    // generic random value from 1 to 100 only for tests so far
    id: Math.floor(Math.random() * 100) + 1,
    userId: req.body.userId,
    quantity: req.body.quantity,
    shipDate: req.body.shipDate,
    status: OrderStatus.Placed,
    complete: false,
  }

  orders.push(order)

  return formatOutput(res, order, 201, 'order')
}

export let removeOrder = (req: Request, res: Response, next: NextFunction)
=> {
  const id = Number(req.params.id)
  const orderIndex = orders.findIndex(item => item.id === id)

  if (orderIndex === -1) {
    return res.status(404).send()
  }

  orders = orders.filter(item => item.id !== id)

  return formatOutput(res, {}, 204)
}

export let getInventory = (req: Request, res: Response, next: NextFunction)
=> {
  const status = req.query.status
  let inventoryOrders = orders
  if (status) {
    inventoryOrders = inventoryOrders.filter(item => item.status ===
status)
  }

  const grouppedOrders = _.groupBy(inventoryOrders, 'userId')

  return formatOutput(res, grouppedOrders, 200, 'inventory')
}
```

After starting the application, creating a new order, and calling the GET/{id} method, you should see the following output:

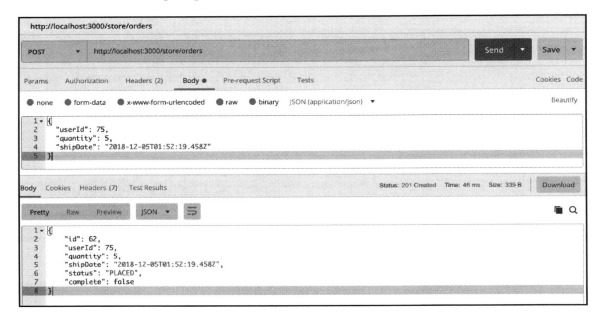

Creating a new order with an ID of 62

Then, you can call the `order`, like so:

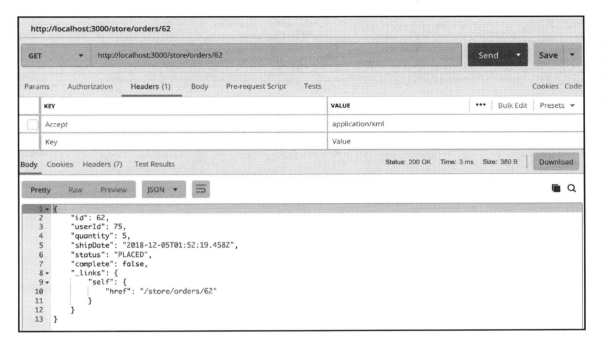

The order with the link refereeing itself

The output of the GET request is in the form of an XML file, as follows:

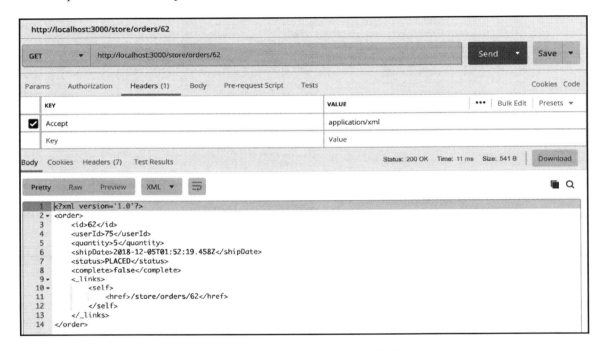

The same information, even though the output is an XML file

Since we know how to use `halson`, let's apply the changes to all operations in the `user` and `order` controllers.

The changes for the `user` controller are shown in the following code snippet:

```
import { NextFunction, Request, Response } from 'express'
import * as halson from 'halson'
import { default as User } from '../models/User'
import { formatOutput } from '../utility/orderApiUtility'

let users: Array<User> = []

export let getUser = (req: Request, res: Response, next: NextFunction) => {
  const username = req.params.username
  let user = users.find(obj => obj.username === username)
  const httpStatusCode = user ? 200 : 404
  if(user){
    user = halson(user).addLink('self', `/users/${user.id}`)
  }
```

```
    return formatOutput(res, user, httpStatusCode, 'user')
}

export let addUser = (req: Request, res: Response, next: NextFunction) => {
  let user: User = {
  // generic random value from 1 to 100 only for tests so far
  id: Math.floor(Math.random() * 100) + 1,
  username: req.body.username,
  firstName: req.body.firstName,
  lastName: req.body.lastName,
  email: req.body.email,
  password: req.body.password,
  phone: req.body.phone,
  userStatus: 1,
  }
  users.push(user)
  user = halson(user).addLink('self', `/users/${user.id}`)
  return formatOutput(res, user, 201, 'user')
}

export let updateUser = (req: Request, res: Response, next: NextFunction)
=> {
  const username = req.params.username
  const userIndex = users.findIndex(item => item.username === username)

  if (userIndex === -1) {
    return res.status(404).send()
  }

  const user = users[userIndex]
  user.username = req.body.username || user.username
  user.firstName = req.body.firstName || user.firstName
  user.lastName = req.body.lastName || user.lastName
  user.email = req.body.email || user.email
  user.password = req.body.password || user.password
  user.phone = req.body.phone || user.phone
  user.userStatus = req.body.userStatus || user.userStatus

  users[userIndex] = user
  return formatOutput(res, {}, 204)
}

export let removeUser = (req: Request, res: Response, next: NextFunction)
=> {
  const username = req.params.username
  const userIndex = users.findIndex(item => item.username === username)

  if (userIndex === -1) {
```

```
    return res.status(404).send()
  }

  users = users.filter(item => item.username !== username)
  return formatOutput(res, {}, 204)
}
```

The changes for the `client` controller are shown in the following code snippet:

```
import { NextFunction, Request, Response } from 'express'
import * as halson from 'halson'
import * as _ from 'lodash'
import { default as Order } from '../models/order'
import { OrderStatus } from '../models/orderStatus'
import { formatOutput } from '../utility/orderApiUtility'

let orders: Array<Order> = []

export let getOrder = (req: Request, res: Response, next: NextFunction) =>
{
  const id = req.params.id
  let order = orders.find(obj => obj.id === Number(id))
  const httpStatusCode = order ? 200 : 404
  if (order) {
 order = halson(order).addLink('self', `/store/orders/${order.id}`)
 }
  return formatOutput(res, order, httpStatusCode, 'order')
}

export let getAllOrders = (req: Request, res: Response, next: NextFunction)
=> {
  const limit = req.query.limit || orders.length
  const offset = req.query.offset || 0

  let filteredOrders = _(orders)
 .drop(offset)
 .take(limit)
 .value()

  filteredOrders = filteredOrders.map(order => {
    return halson(order)
      .addLink('self', `/store/orders/${order.id}`)
      .addLink('user', {
        href: `/users/${order.userId}`,
      })
  })

  return formatOutput(res, filteredOrders, 200, 'order')
```

```
}

export let addOrder = (req: Request, res: Response, next: NextFunction) =>
{
  let order: Order = {
    // generic random value from 1 to 100 only for tests so far
    id: Math.floor(Math.random() * 100) + 1,
    userId: req.body.userId,
    quantity: req.body.quantity,
    shipDate: req.body.shipDate,
    status: OrderStatus.Placed,
    complete: false,
  }

  orders.push(order)
  order = halson(order)
    .addLink('self', `/store/orders/${order.id}`)
    .addLink('user', {
      href: `/users/${order.userId}`,
    })
  return formatOutput(res, order, 201, 'order')
}

export let removeOrder = (req: Request, res: Response, next: NextFunction)
=> {
  const id = Number(req.params.id)
  const orderIndex = orders.findIndex(item => item.id === id)

  if (orderIndex === -1) {
    return res.status(404).send()
  }

  orders = orders.filter(item => item.id !== id)

  return formatOutput(res, {}, 204)
}

export let getInventory = (req: Request, res: Response, next: NextFunction)
=> {
  const status = req.query.status
  let inventoryOrders = orders
  if (status) {
    inventoryOrders = inventoryOrders.filter(item => item.status ===
status)
  }

  const grouppedOrders = _.groupBy(inventoryOrders, 'userId')
```

```
    return formatOutput(res, grouppedOrders, 200, 'inventory')
}
```

If you create a new user, you should see the self-relationship, as follows:

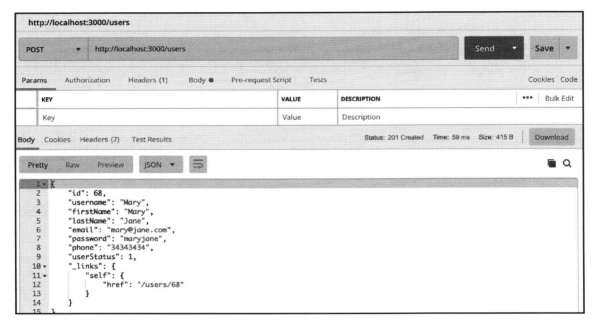

Self-link when a new user is created

This will also happen when you get a new user:

Getting the user with a self-link

The same approach applies to the orders, but we also have to include the user as an external reference:

New order being created and the output showing the user relationship besides the order itself

This also works with arrays and XML:

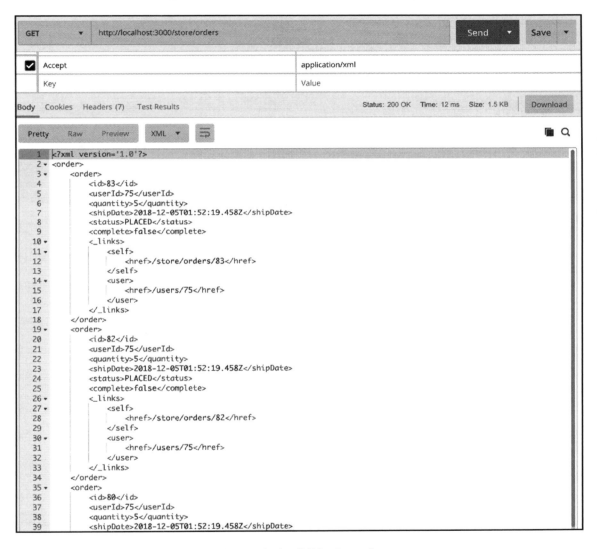

HAL in an array situation with XML as the output format

Summary

This chapter walked you through some techniques to help you with developing applications where you need to specify the content type definition and make sure that the server is prepared to handle different content mime types. If it is not supported, at least you know how to set the correct an HTTP status code, should it be retrieved by the requester.

You also had the opportunity to apply XML and JSON object conversions, as well as including HAL in the `order-api` application.

The next `Chapter 8`, *Working with Databases and ODMs,* is going to go a little bit further and replace our in-memory storage with MongoDB.

Questions

1. What do we mean by content negotiation?
2. Is it required that an application validates the `Accept` request?
3. What HTTP status code should be sent if the client sends an unsupported mime-type in the Accept header?
4. What does HAL mean?
5. Is it possible to use HAL with XML?
6. Do you have to expose all resources as JSON and XML?
7. Is `halson` the only package that can be used for HAL with TypeScript?

Further reading

To improve your knowledge in regard to formatting strategy, the following book is recommended, as it will be helpful for the upcoming chapters:

- *RESTful Web API Design with Node.js* (`https://www.packtpub.com/web-development/restful-web-api-design-nodejs-10-third-edition`)

3
Section 3: Enhancing RESTful Web Services

In this section, you will learn how to enhance the capabilities of your RESTful web services with database interactions, securing, and testing, as well as error handling for debugging.

The following chapters are included in this section:

- Chapter 8, *Working with Databases and ODMs*
- Chapter 9, *Securing your API*
- Chapter 10, *Error Handling and Logging*
- Chapter 11, *Creating CI/CD Pipelines for Your API*

Working with Databases and ODMs

8

Having persistent data is a key requirement for every web service. At any given time, the web service should return/serve the same data to simultaneous API calls. In this chapter, we will learn how to set up a MongoDB server and connect it to your API. We will create some simple database wrapper methods that will help our internal logic while not mixing it with an external dependency.

The following topics will be covered in this chapter:

- Using MongoDB
- Setting up `order-api` with MongoDB
- Manual testing

Technical requirements

All information required to run the code from this chapter is provided by this chapter itself. The only necessity is to get the previous installation process from the previous chapter done, such as NodeJS, VS Code, and TypeScript.

All code bases are available at `https://github.com/PacktPublishing/Hands-On-RESTful-Web-Services-with-TypeScript-3/tree/master/Chapter08`.

Using MongoDB

In a few words, MongoDB is a database ruled by the document-oriented concept. It is currently the most popular NoSQL database and it is open source. Since 2015, and as of 2019, MongoDB ranks on top of popular databases. *10gen* had developed it initially, and it has evolved to be known as MongoDB Inc.

As a document-oriented database, MongoDB stores data in JSON documents with a dynamic schema. So, when the records have to be stored, it can be done irrespective of the data structure. The documents in MongoDB bear resemblance to JSON objects, as shown:

```
{
 "firstName" : "Biharck",
 "lastName": "Araujo",
 "phone" : "5555555",
 "username" : "biharck"
}
```

Installing MongoDB

If you want to install MongoDB locally, perform the following steps:

1. You can go to the MongoDB website at `https://www.mongodb.com/`, go to the download section, select the community version, and download it:

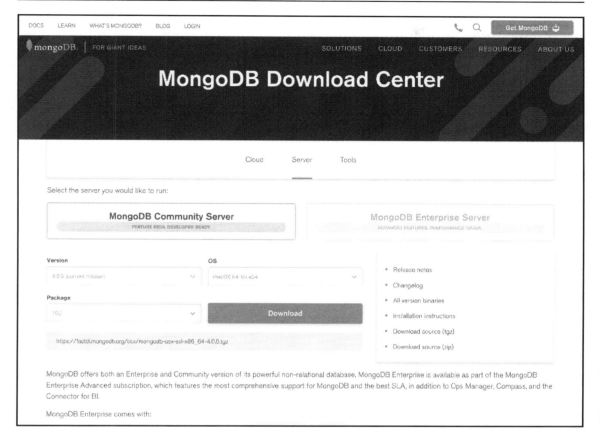

MongoDB download section

 You can always get the latest installation process through the MongoDB documentation website: `https://docs.mongodb.com/manual/tutorial/` `install-mongodb-on-os-x/`.

2. The installation process is really easy. Once you have downloaded the file, extract the file with the following command:

```
$ tar -zxvf mongodb-osx-ssl-x86_64-4.0.5.tgz
```

The filename might change due to the version you're using.

As explained on the MongoDB documentation, the `bin/` directory of the `tar` ball contains the MongoDB binaries. You need to perform any of the following actions:

- Find the directory listed in your `PATH` variable, such as `/usr/local/bin`, and copy these binary files to that directory
- Again, from a directory listed in your `PATH` variable, create `symbolic` links to each of these binaries
- The user's `PATH` environment variable has to be modified to include this directory

3. After that, you are able to run the MongoDB. In order to do that, you have to create a data directory, where the content will be stored:

```
$ mkdir -p /data/db
```

Do not forget to add the permissions to the `/data/db` folder, so that mongo will be able to write and read files there.

4. Now, you can run MongoDB with the following command:

```
$ mongod
```

5. And you can check whether MongoDB is running using the `mongo` client. To use it, you just have to type the following:

```
$ mongo
```

6. Use the preceding command on the same machine as `mongod` is running and the output might be similar to this:

```
# mongo
MongoDB shell version v3.6.3
connecting to: mongodb://127.0.0.1:27017
MongoDB server version: 3.6.3
Welcome to the MongoDB shell.
For interactive help, type "help".
```

```
For more comprehensive documentation, see
        http://docs.mongodb.org/
Questions? Try the support group
        http://groups.google.com/group/mongodb-user
Server has startup warnings:
2018-12-24T03:32:01.107+0000 I CONTROL  [initandlisten]
2018-12-24T03:32:01.107+0000 I CONTROL  [initandlisten] ** WARNING:
Access control is not enabled for the database.
2018-12-24T03:32:01.107+0000 I CONTROL  [initandlisten] ** Read and
write access to data and configuration is unrestricted.
2018-12-24T03:32:01.107+0000 I CONTROL  [initandlisten]
>
```

7. Then, run the following command to list the available databases:

```
$ show databases
```

The output must be as follows:

```
> show databases
admin 0.000GB
local 0.000GB
order-api 0.000GB
ppi 0.005GB
test 0.000GB
```

This completes our MongoDB installation. Let's see how it works with Docker.

MongoDB with Docker

Another possibility, and the one we are going to employ from now on, is using MongoDB through Docker.

In a nutshell, Docker is an open source platform, developed in the Go language and created by Google. Because of its high performance, the software ensures greater ease when creating and administering isolated environments, thereby ensuring faster availability of programs for the end user.

As you can see in the following diagram, the OS is a layer shared among the applications, and the container engine layer is responsible for orchestrating that. The idea is to have isolated resources rather than the OS:

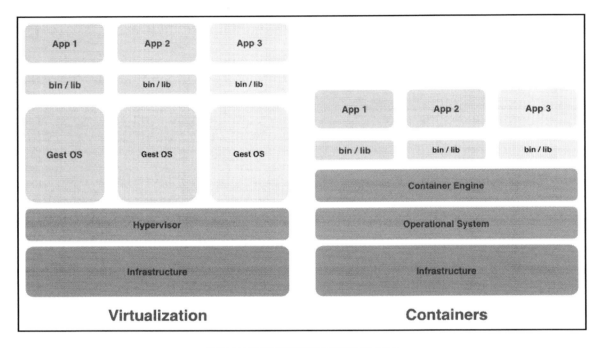

Comparison between machine Virtualization and Containers

 Do you want to learn more about Docker? Check out their documentation website: https://docs.docker.com/.

Installing Docker

The installation process is simple. Perform the following steps:

1. Go to their download website, https://www.docker.com/get-started, and download the version based on your operating system:

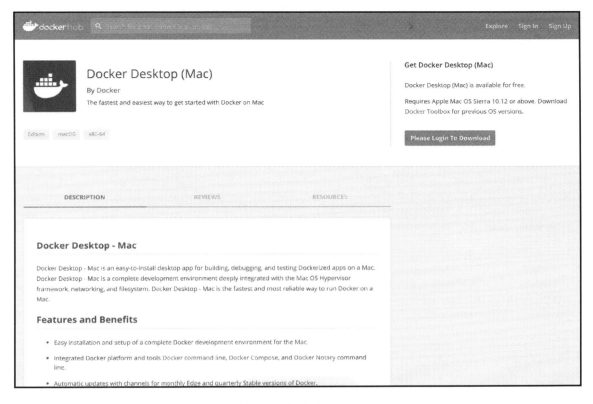

Docker download section for Mac users

2. After you download the proper version, double-click `Docker.dmg` to start the installation process. When the installation completes and Docker starts, the Docker icon will appear in the status bar:

Docker icon in status bar for Mac users

3. You can also go to the Terminal and check the `docker version` with the following command:

```
$ docker version
```

The output will be similar to the following:

```
Client: Docker Engine - Community
 Version: 18.09.0
 API version: 1.39
 Go version: go1.10.4
 Git commit: 4d60db4
 Built: Wed Nov 7 00:47:43 2018
 OS/Arch: darwin/amd64
 Experimental: false

Server: Docker Engine - Community
 Engine:
  Version: 18.09.0
  API version: 1.39 (minimum version 1.12)
  Go version: go1.10.4
  Git commit: 4d60db4
  Built: Wed Nov 7 00:55:00 2018
  OS/Arch: linux/amd64
  Experimental: true
```

Of course, this information may change due to the OS, Docker version, and so on.

Running MongoDB on Docker

Now that we have Docker running, we can start a new MongoDB process through Docker. This process is really easy:

1. Go to the Terminal and type the following:

   ```
   $ docker run --name my-mongo -p 27017:27017 -d mongo:latest
   ```

   Every time you restart Docker, the content will be lost. To avoid that, you can create a volume on the host machine.

2. Create the following volume on the host machine:

   ```
   $ docker run --name my-mongo -p 27017:27017 -v /data/mongo:/data/db
   -d mongo:latest
   ```

   This command will create a new MongoDB through Docker. The container's name will be my-mongo, the port will be mapped to 27017, and we will get the latest docker image at https://hub.docker.com/_/mongo/.

3. Once this command finishes, type the following command:

   $ docker ps

   You should be able to see the container running:

Listing the new container created using the ps command

4. Now, let's go to this container and see whether MongoDB is running. To do so, run the `exec` command, as follows, to go to the container we created:

   $ docker exec -it my-mongo sh

5. Then, run the following command to connect to the MongoDB:

   $ mongo

The output should look as follows:

```
MongoDB shell version v3.6.3
connecting to: mongodb://127.0.0.1:27017
MongoDB server version: 3.6.3
Welcome to the MongoDB shell.
For interactive help, type "help".
For more comprehensive documentation, see
  http://docs.mongodb.org/
Questions? Try the support group
  http://groups.google.com/group/mongodb-user
Server has startup warnings:
2018-12-27T01:56:17.455+0000 I STORAGE [initandlisten]
2018-12-27T01:56:17.456+0000 I STORAGE [initandlisten] ** WARNING: Using
the XFS filesystem is strongly recommended with the WiredTiger storage
engine
2018-12-27T01:56:17.456+0000 I STORAGE [initandlisten] ** See
http://dochub.mongodb.org/core/prodnotes-filesystem
2018-12-27T01:56:18.144+0000 I CONTROL [initandlisten]
2018-12-27T01:56:18.145+0000 I CONTROL [initandlisten] ** WARNING: Access
control is not enabled for the database.
2018-12-27T01:56:18.145+0000 I CONTROL [initandlisten] ** Read and write
access to data and configuration is unrestricted.
2018-12-27T01:56:18.145+0000 I CONTROL [initandlisten]
```

If you see the previous output, everything is fine and we are good to move on to the next section.

Robomongo

Now that we have MongoDB running through Docker, we will install a MongoDB client to make our life easier when we want to manipulate the documents stored on MongoDB.

There are a lot of tools you can use to do that. One of them is Robomongo, which is free and easy to use. To use it, you need to go through the following steps:

1. Visit `https://robomongo.org/` and download Robomongo. The page looks as follows:

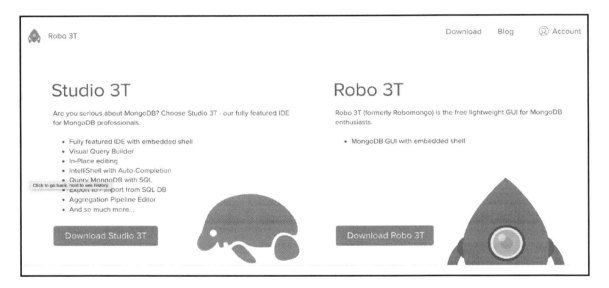

Robomongo web page

2. Once you download Robomongo, double-click on it and the configuration section will appear:

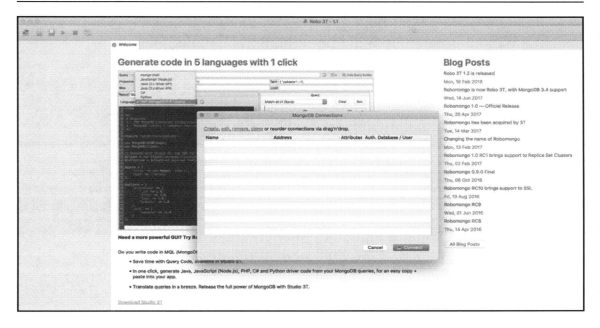

The main page on the Robomongo client

3. Click on **Create a section** to create a new connection with Docker:

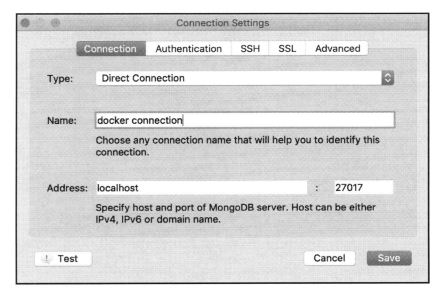

Connection settings section

4. Fill in the name as `docker connection` and the address as `localhost`. Keep the port as `27017` and type `test` on the bottom-left side:

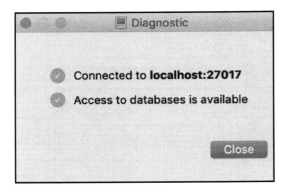

Testing connection successful

You will see the new connection created as shown:

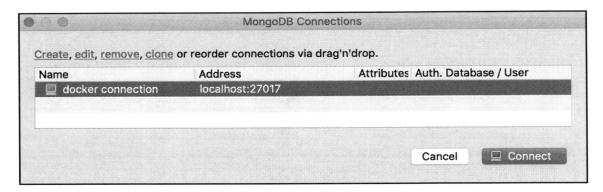

New connection created

5. Click on **Save and Connect**. You should see the connection established, as shown here:

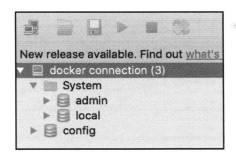

Connection tree

Since we have MongoDB already configured, it is time to start implementing the communication with the `order-api` application. The changes will be shown in the following sections.

Setting up order-api with Mongo

In this section, I will walk you through the changes on the `order-api` application so that we can start using a database to store and retrieve information rather than in-memory.

We will use a library called `mongoose` to help us. Mongoose is an **Object Data Modeling (ODM)** library that provides highly rigorous modeling for your data, enforcing structure as needed, while still maintaining the flexibility that makes MongoDB powerful, such as a document-oriented database.

Configuring Mongoose

Configuration is easy, just like the other libraries used in the previous sections with npm.

1. Start by running the following command:

   ```
   $ npm install mongoose --save
   ```

2. Also install the TypeScript @types:

   ```
   $ npm install @types/mongodb @types/mongoose --save-dev
   ```

So far, our `package.json` file should look as follows:

```json
{
  "name": "order-api",
  "version": "1.0.0",
  "description": "This is the example from the Book Hands on RESTful
APIs with TypeScript 3",
  "main": "./dist/server.js",
  "scripts": {
    "build": "npm run clean && tsc",
    "clean": "rimraf dist && rimraf reports",
    "lint": "tslint ./src/**/*.ts ./test/**/*.spec.ts",
    "lint:fix": "tslint --fix ./src/**/*.ts ./test/**/*.spec.ts -t
verbose",
    "pretest": "cross-env NODE_ENV=test npm run build && npm run lint",
    "test": "cross-env NODE_ENV=test mocha --reporter spec --compilers
ts:ts-node/register test/**/*.spec.ts --exit",
    "test:mutation": "stryker run",
    "stryker:init": "stryker init",
    "dev": "cross-env PORT=3000 NODE_ENV=dev ts-node ./src/server.ts",
    "prod": "PORT=3000 npm run build && npm run start",
    "tsc": "tsc"
  },
  "engines": {
    "node": ">=8.0.0"
  },
  "keywords": [
    "order POC",
    "Hands on RESTful APIs with TypeScript 3",
    "TypeScript 3",
    "Packt Books"
  ],
  "author": "Biharck Muniz Araújo",
  "license": "MIT",
  "devDependencies": {
    "@types/body-parser": "^1.17.0",
    "@types/chai": "^4.1.7",
    "@types/chai-http": "^3.0.5",
    "@types/express": "^4.16.0",
    "@types/mocha": "^5.2.5",
    "@types/mongodb": "^3.1.17",
    "@types/mongoose": "^5.3.5",
    "@types/node": "^10.12.12",
    "chai": "^4.2.0",
    "cross-env": "^5.2.0",
    "mocha": "^5.2.0",
    "rimraf": "^2.6.2",
    "stryker": "^0.33.1",
```

```
            "stryker-api": "^0.22.0",
            "stryker-html-reporter": "^0.16.9",
            "stryker-mocha-framework": "^0.13.2",
            "stryker-mocha-runner": "^0.15.2",
            "stryker-typescript": "^0.16.1",
            "ts-node": "^7.0.1",
            "tslint": "^5.11.0",
            "tslint-config-prettier": "^1.17.0",
            "typescript": "^3.2.1"
        },
        "dependencies": {
            "body-parser": "^1.18.3",
            "express": "^4.16.4",
            "halson": "^3.0.0",
            "js2xmlparser": "^3.0.0",
            "lodash": "^4.17.11",
            "mongoose": "^5.4.0"
        }
    }
```

With Mongoose configured, we will move on to create schemas.

Creating the schemas

So far, we've created the models based on the TypeScript interface and we are using them to persist the information in memory. With Mongoose, we have to introduce the schemas concept, which basically does the mapping between the code base and the MongoDB document.

Due to our two models, user and order, we must create two new schemas, also called user and order, under the src/schemas folder. The src/schemas/user.ts content is as follows:

```
import { Document, Model, model, Schema } from 'mongoose'
import { default as User } from '../models/user'

export interface UserModel extends User, Document {}

export const UserSchema: Schema = new Schema({
  firstName: String,
  lastName: String,
  email: String,
  password: String,
  phone: String,
  userStatus: Number,
```

```
    username: String,
})

export const UserModel: Model<UserModel> = model<UserModel>('User',
UserSchema)
```

Note that we are now extending the user model and document, which is a MongoDB document mapped by Mongoose, and defining a new mongo schema called UserModel.

 By definition, all resources using Mongoose have a property called _id, which is added to all MongoDB documents by default, and have a default type of ObjectId.

The same idea applies to the src/schemas/order.ts file:

```
import { Document, Model, model, Schema } from 'mongoose'
import { default as Order } from '../models/order'
import { OrderStatus } from '../models/orderStatus'

export interface OrderModel extends Order, Document {}

export const OrderSchema: Schema = new Schema({
  userId: { type: Schema.Types.ObjectId, ref: 'User' },
  quantity: Number,
  shipDate: Date,
  status: { type: String, enum: ['PLACED', 'APPROVED', 'DELIVERED'] },
  complete: Boolean,
})

export const OrderModel: Model<OrderModel> = model<OrderModel>(
  'Order',
  OrderSchema
)
```

Note that OrderSchema is referring to the userId directly to the schema user:

```
...
userId: { type: Schema.Types.ObjectId, ref: 'User' },
...
```

So, it is now required that the user exists before we create a new order.

Changing tests

Since we are no longer using an in-memory strategy to persist the information, we have to change our tests to be compliant with the new approach.

 For this book's purpose, we are using integration tests and mutation tests. There are a lot of different strategies to test your application, such as unit testing and **Behavior-Driven Design (BDD)**. Check out the *Further reading* section for a list of books that you can consult to improve your testing skills.

Since our tests depend on a sequence, we will rename them using a number as a prefix to guarantee that they will be executed in the order we desire. This is only one way to do that. There are a lot of different possibilities you can try out:

Renaming the tests by adding a number version in front of them

The next step is to change the `02.user.spec.ts` content as follows:

```
'use strict'

import * as chai from 'chai'
import chaiHttp = require('chai-http')
import 'mocha'
import app from '../../src/app'
import { UserModel } from '../../src/schemas/User'

chai.use(chaiHttp)

const expect = chai.expect

const user = {
  username: 'John',
  firstName: 'John',
  lastName: 'Doe',
  email: 'John@myemail.com',
  password: 'password',
  phone: '5555555',
  userStatus: 1,
```

```
}

describe('userRoute', () => {
  before(async () => {
    expect(UserModel.modelName).to.be.equal('User')
    UserModel.collection.drop()
  })
  it('should respond with HTTP 404 status because there is no user', async
() => {
    return chai
      .request(app)
      .get(`/users/${user.username}`)
      .then(res => {
        expect(res.status).to.be.equal(404)
      })
  })
  it('should create a new user and retrieve it back', async () => {
    return chai
      .request(app)
      .post('/users')
      .send(user)
      .then(res => {
        expect(res.status).to.be.equal(201)
        expect(res.body.username).to.be.equal(user.username)
      })
  })
  it('should return the user created on the step before', async () => {
    return chai
      .request(app)
      .get(`/users/${user.username}`)
      .then(res => {
        expect(res.status).to.be.equal(200)
        expect(res.body.username).to.be.equal(user.username)
      })
  })
  it('should updated the user John', async () => {
    user.username = 'John_Updated'
    user.firstName = 'John Updated'
    user.lastName = 'Doe Updated'
    user.email = 'John@myemail_updated.com'
    user.password = 'password Updated'
    user.phone = '3333333'
    user.userStatus = 12

    return chai
      .request(app)
      .patch(`/users/John`)
      .send(user)
```

```
          .then(res => {
            expect(res.status).to.be.equal(204)
          })
    })
  it('should return the user updated on the step before', async () => {
      return chai
        .request(app)
        .get(`/users/${user.username}`)
        .then(res => {
          expect(res.status).to.be.equal(200)
          expect(res.body.username).to.be.equal(user.username)
          expect(res.body.firstName).to.be.equal(user.firstName)
          expect(res.body.lastName).to.be.equal(user.lastName)
          expect(res.body.email).to.be.equal(user.email)
          expect(res.body.password).to.be.equal(user.password)
          expect(res.body.phone).to.be.equal(user.phone)
          expect(res.body.userStatus).to.be.equal(user.userStatus)
        })
    })
  it('should return 404 because the user does not exist', async () => {
      user.firstName = 'Mary Jane'

      return chai
        .request(app)
        .patch(`/users/Mary`)
        .send(user)
        .then(res => {
          expect(res.status).to.be.equal(404)
        })
    })
  it('should remove an existent user', async () => {
      return chai
        .request(app)
        .del(`/users/${user.username}`)
        .then(res => {
          expect(res.status).to.be.equal(204)
        })
    })
  it('should return 404 when it is trying to remove an user because the
user does not exist', async () => {
      return chai
        .request(app)
        .del(`/users/Mary`)
        .then(res => {
          expect(res.status).to.be.equal(404)
        })
    })
})
```

There are a few changes you might have noticed, including the following:

- There is a new import to the `user` schema:

```
import { UserModel } from '../../src/schemas/User'
```

- We are now defining a simple JavaScript object for the `user` request:

```
const user = {
  username: 'John',
  firstName: 'John',
  lastName: 'Doe',
  email: 'John@myemail.com',
  password: 'password',
  phone: '5555555',
  userStatus: 1,
}
```

- There is a new section called `before`, which runs once every time prior to starting the `user` tests. This new section checks whether the model is `user` and drops the collections if exists in order to do not mess up with the tests:

```
before(async () => {
  expect(UserModel.modelName).to.be.equal('User')
  UserModel.collection.drop()
})
```

And the same idea applies to the `03_order.spec.ts` file:

```
'use strict'

import * as chai from 'chai'
import chaiHttp = require('chai-http')
import 'mocha'
import * as mongoose from 'mongoose'
import app from '../../src/app'
import { OrderStatus } from '../../src/models/orderStatus'
import { OrderModel } from '../../src/schemas/order'

chai.use(chaiHttp)

const expect = chai.expect

const order = {
  userId: 20,
  quantity: 1,
  shipDate: new Date(),
  status: OrderStatus.Placed,
```

```
    complete: false,
}

let orderIdCreated

describe('userRoute', () => {
  before(async () => {
    expect(OrderModel.modelName).to.be.equal('Order')
    OrderModel.collection.drop()
  })

  it('should respond with HTTP 404 status because there is no order', async
() => {
    return chai
      .request(app)
      .get(`/store/orders/000`)
      .then(res => {
        expect(res.status).to.be.equal(404)
      })
  })

  it('should create a new user for Order tests and retrieve it back', async
() => {
    const user = {
      username: 'OrderUser',
      firstName: 'Order',
      lastName: 'User',
      email: 'order@myemail.com',
      password: 'password',
      phone: '5555555',
      userStatus: 1,
    }
    return chai
      .request(app)
      .post('/users')
      .send(user)
      .then(res => {
        expect(res.status).to.be.equal(201)
        expect(res.body.username).to.be.equal(user.username)
        order.userId = res.body._id
      })
  })

  it('should create a new order and retrieve it back', async () => {
    return chai
      .request(app)
      .post(`/store/orders`)
      .send(order)
```

```
          .then(res => {
            expect(res.status).to.be.equal(201)
            expect(res.body.userId).to.be.equal(order.userId)
            expect(res.body.complete).to.be.equal(false)
            orderIdCreated = res.body._id
          })
      })
    it('should return the order created on the step before', async () => {
      return chai
        .request(app)
        .get(`/store/orders/${orderIdCreated}`)
        .then(res => {
          expect(res.status).to.be.equal(200)
          expect(res.body._id).to.be.equal(orderIdCreated)
          expect(res.body.status).to.be.equal(order.status)
        })
      })
    it('should return all orders so far', async () => {
      return chai
        .request(app)
        .get(`/store/orders`)
        .then(res => {
          expect(res.status).to.be.equal(200)
          expect(res.body.length).to.be.equal(1)
        })
      })
    it('should not return orders because offset is higher than the size of
the orders array', async () => {
      return chai
        .request(app)
        .get(`/store/orders?offset=2&limit=2`)
        .then(res => {
          expect(res.status).to.be.equal(200)
          expect(res.body.length).to.be.equal(0)
        })
      })
    it('should return the inventory for all users', async () => {
      return chai
        .request(app)
        .get(`/store/inventory?status=PLACED`)
        .then(res => {
          expect(res.status).to.be.equal(200)
          expect(res.body[order.userId].length).to.be.equal(1)
        })
      })
    it('should remove an existing order', async () => {
      return chai
        .request(app)
```

```
    .del(`/store/orders/${orderIdCreated}`)
    .then(res => {
      expect(res.status).to.be.equal(204)
    })
  })
  it('should return 404 when it is trying to remove an order because the
order does not exist', async () => {
    return chai
      .request(app)
      .del(`/store/orders/${orderIdCreated}`)
      .then(res => {
        expect(res.status).to.be.equal(404)
      })
  })
})
})
```

- There is a new import to the OrderSchema:

```
import { OrderStatus } from '../../src/models/orderStatus'
```

- The request order is now a simple JavaScript object:

```
const order = {
  userId: 20,
  quantity: 1,
  shipDate: new Date(),
  status: OrderStatus.Placed,
  complete: false,
}
```

- There is a new variable called orderIdCreated, which is going to store the orderID when created:

```
let orderIdCreated
```

- As user.spec, there is a new step that runs once every time that the order test runs, which cleans up the order collection and checks whether the model's name is Order:

```
before(async () => {
  expect(OrderModel.modelName).to.be.equal('Order')
  OrderModel.collection.drop()
})
```

- There is a new test that creates a new user to be used on the order tests:

```
it('should create a new user for Order tests and retrieve it back',
async () => {
```

```
            const user = {
              username: 'OrderUser',
              firstName: 'Order',
              lastName: 'User',
              email: 'order@myemail.com',
              password: 'password',
              phone: '5555555',
              userStatus: 1,
            }
            return chai
              .request(app)
              .post('/users')
              .send(user)
              .then(res => {
                expect(res.status).to.be.equal(201)
                expect(res.body.username).to.be.equal(user.username)
                order.userId = res.body._id
              })
          })
```

The rest is essentially the same except for possible improvements.

Of course, if you run the tests, will you see a lot of errors, which is expected as we haven't changed the controllers yet. Let's walk through the changes in the following sections.

Changing the controllers

Since we now have the tests in place and failing, we can change our user and order controllers to start using MongoDB. First, let's change the user controller. The user controller content is as follows:

```
import { NextFunction, Request, Response } from 'express'
import * as halson from 'halson'
import { UserModel } from '../schemas/User'
import { formatOutput } from '../utility/orderApiUtility'

export let getUser = (req: Request, res: Response, next: NextFunction) => {
  const username = req.params.username

  UserModel.findOne({ username: username }, (err, user) => {
    if (!user) {
      return res.status(404).send()
    }

    user = user.toJSON()
    user._id = user._id.toString()
```

```
    user = halson(user).addLink('self', `/users/${user._id}`)
    return formatOutput(res, user, 200, 'user')
  })
}

export let addUser = (req: Request, res: Response, next: NextFunction) => {
  const newUser = new UserModel(req.body)

  newUser.save((error, user) => {
    user = halson(user.toJSON()).addLink('self', `/users/${user._id}`)
    return formatOutput(res, user, 201, 'user')
  })
}

export let updateUser = (req: Request, res: Response, next: NextFunction)
=> {
  const username = req.params.username

  UserModel.findOne({ username: username }, (err, user) => {
    if (!user) {
      return res.status(404).send()
    }

    user.username = req.body.username || user.username
    user.firstName = req.body.firstName || user.firstName
    user.lastName = req.body.lastName || user.lastName
    user.email = req.body.email || user.email
    user.password = req.body.password || user.password
    user.phone = req.body.phone || user.phone
    user.userStatus = req.body.userStatus || user.userStatus

    user.save(error => {
      res.status(204).send()
    })
  })
}

export let removeUser = (req: Request, res: Response, next: NextFunction)
=> {
  const username = req.params.username

  UserModel.findOne({ username: username }, (err, user) => {
    if (!user) {
      return res.status(404).send()
    }

    user.remove(error => {
      res.status(204).send()
```

```
    })
  })
}
```

Let's walk through this file:

- Notice that now, we import the schema:

```
import { UserModel } from '../schemas/User'
```

- Now, we are using the find method from mongoose to get the User document directly to MongoDB instead of getting the information from an array, as before. We are still using the username as a filter and getting it as a parameter:

```
export let getUser = (req: Request, res: Response, next:
NextFunction) => {
  const username = req.params.username

  UserModel.findOne({ username: username }, (err, user) => {
    if (!user) {
      return res.status(404).send()
    }

    user = user.toJSON()
    user._id = user._id.toString()

    user = halson(user).addLink('self', `/users/${user._id}`)
    return formatOutput(res, user, 200, 'user')
  })
}
```

- The same idea applies to the addUser method, except we use the save method:

```
export let addUser = (req: Request, res: Response, next:
NextFunction) => {
  const newUser = new UserModel(req.body)

  newUser.save((error, user) => {
    user = halson(user.toJSON()).addLink('self',
`/users/${user._id}`)
    return formatOutput(res, user, 201, 'user')
  })
}
```

- The updated User is a bit more complex, even though it is still fairly simple. The difference is that first, we search for the user and with its document, we do the changes and call the save method again:

```
export let updateUser = (req: Request, res: Response, next:
NextFunction) => {
  const username = req.params.username

  UserModel.findOne({ username: username }, (err, user) => {
    if (!user) {
      return res.status(404).send()
    }

    user.username = req.body.username || user.username
    user.firstName = req.body.firstName || user.firstName
    user.lastName = req.body.lastName || user.lastName
    user.email = req.body.email || user.email
    user.password = req.body.password || user.password
    user.phone = req.body.phone || user.phone
    user.userStatus = req.body.userStatus || user.userStatus

    user.save(error => {
      res.status(204).send()
    })
  })
}
```

- The same idea applies to the remove operation. First, we search for the User, and then, with the user document, we call the remove method:

```
export let removeUser = (req: Request, res: Response, next:
NextFunction) => {
  const username = req.params.username

  UserModel.findOne({ username: username }, (err, user) => {
    if (!user) {
      return res.status(404).send()
    }

    user.remove(error => {
      res.status(204).send()
    })
  })
}
```

In general words, the order controller changed almost like the `User` controller such as the following file:

```
import { NextFunction, Request, Response } from 'express'
import * as halson from 'halson'
import * as _ from 'lodash'
import { OrderModel } from '../schemas/order'
import { UserModel } from '../schemas/User'
import { formatOutput } from '../utility/orderApiUtility'

export let getOrder = (req: Request, res: Response, next: NextFunction) =>
{
  const id = req.params.id
  OrderModel.findById(id, (err, order) => {
    if (!order) {
      return res.status(404).send()
    }
    order = halson(order.toJSON()).addLink('self',
`/store/orders/${order.id}`)
    return formatOutput(res, order, 200, 'order')
  })
}

export let getAllOrders = (req: Request, res: Response, next: NextFunction)
=> {
  const limit = Number(req.query.limit) || 0
  const offset = Number(req.query.offset) || 0

  OrderModel.find({}, null, { skip: offset, limit: limit }).then(orders =>
{
    if (orders) {
      orders = orders.map(order => {
        return halson(order.toJSON())
          .addLink('self', `/store/orders/${order.id}`)
          .addLink('user', {
            href: `/users/${order.userId}`,
          })
      })
    }
    return formatOutput(res, orders, 200, 'order')
  })
}

export let addOrder = (req: Request, res: Response, next: NextFunction) =>
{
  const userId = req.body.userId

  UserModel.findById(userId, (err, user) => {
```

```
    if (!user) {
      return res.status(404).send()
    }

    const newOrder = new OrderModel(req.body)

    newOrder.save((error, order) => {
      order = halson(order.toJSON())
        .addLink('self', `/store/orders/${order._id}`)
        .addLink('user', {
          href: `/users/${order.userId}`,
        })

      return formatOutput(res, order, 201, 'order')
    })
  })
}

export let removeOrder = (req: Request, res: Response, next: NextFunction)
=> {
  const id = req.params.id
  OrderModel.findById(id, (err, order) => {
    if (!order) {
      return res.status(404).send()
    }
    order.remove(error => {
      res.status(204).send()
    })
  })
}

export let getInventory = (req: Request, res: Response, next: NextFunction)
=> {
  const status = req.query.status
  OrderModel.find({ status: status }, (err, orders) => {
    orders = _.groupBy(orders, 'userId')
    return formatOutput(res, orders, 200, 'inventory')
  })
}
```

- There is a new import for the Order schema:

```
import { OrderModel } from '../schemas/order'
```

- Also to the User schema:

```
import { UserModel } from '../schemas/User'
```

- The `get` order uses the `findById` method from mongoose, using the order `id` as a parameter:

```
export let getOrder = (req: Request, res: Response, next:
NextFunction) => {
  const id = req.params.id
  OrderModel.findById(id, (err, order) => {
    if (!order) {
      return res.status(404).send()
    }
    order = halson(order.toJSON()).addLink('self',
`/store/orders/${order.id}`)
    return formatOutput(res, order, 200, 'order')
  })
}
```

- The `getAllOrders` method does the math for limit and offsets as mongoose parameter:

```
export let getAllOrders = (req: Request, res: Response, next:
NextFunction) => {
  const limit = Number(req.query.limit) || 0
  const offset = Number(req.query.offset) || 0

  OrderModel.find({}, null, { skip: offset, limit: limit
}).then(orders => {
    if (orders) {
      orders = orders.map(order => {
        return halson(order.toJSON())
          .addLink('self', `/store/orders/${order.id}`)
          .addLink('user', {
            href: `/users/${order.userId}`,
          })
      })
    }
    return formatOutput(res, orders, 200, 'order')
  })
}
```

- `addOrder` first gets the `user`, and then saves the `order`:

```
export let addOrder = (req: Request, res: Response, next:
NextFunction) => {
  const userId = req.body.userId

  UserModel.findById(userId, (err, user) => {
    if (!user) {
      return res.status(404).send()
```

```
    }

    const newOrder = new OrderModel(req.body)

    newOrder.save((error, order) => {
      order = halson(order.toJSON())
        .addLink('self', `/store/orders/${order._id}`)
        .addLink('user', {
          href: `/users/${order.userId}`,
        })

      return formatOutput(res, order, 201, 'order')
    })
  })
}
```

- remove. order retrieves the order before removing it:

```
export let removeOrder = (req: Request, res: Response, next:
NextFunction) => {
  const id = req.params.id
  OrderModel.findById(id, (err, order) => {
    if (!order) {
      return res.status(404).send()
    }
    order.remove(error => {
      res.status(204).send()
    })
  })
}
```

- The getInventory operation finds the orders and groups them all based on userId:

```
export let getInventory = (req: Request, res: Response, next:
NextFunction) => {
  const status = req.query.status
  OrderModel.find({ status: status }, (err, orders) => {
    orders = _.groupBy(orders, 'userId')
    return formatOutput(res, orders, 200, 'inventory')
  })
}
```

Finally, if we run the tests, they should pass:

```
$ npm run test
```

You will see the following output:

```
baseRoute
  should respond with HTTP 200 status (109ms)
  should respond with success message

userRoute
  should respond with HTTP 404 status because there is no user (79ms)
  should create a new user and retrieve it back (84ms)
  should return the user created on the step before
  should updated the user John
  should return the user updated on the step before
  should return 404 because the user does not exist
  should remove an existent user
  should return 404 when it is trying to remove an user because the user
does not exist

userRoute
  should respond with HTTP 404 status because there is no order
  should create a new user for Order tests and retrieve it back
  should create a new order and retrieve it back (56ms)
  should return the order created on the step before
  should return all orders so far
  should not return orders because offset is higher than the size of the
orders array
  should return the inventory for all users
  should remove an existing order
  should return 404 when it is trying to remove an order because the
order does not exist


19 passing (643ms)
```

The same applies to Stryker:

```
$ stryker run
```

The previous command will show the following output:

Stryker report after mongo changes

Manual testing

Since we have everything in place with MongoDB, it is time to start the application and run some tests manually. We will do so using the following steps:

1. To start the application, run the following command:

```
$ npm run dev
```

2. Go to the POSTman application and create a new user:

POST operation to create a new user

3. Now, try to retrieve it using the GET operation:

Retrieving the user through the GET method

4. Try to create more `users` and then create `orders` for them all:

Creating orders using the POST operation for the new users created

5. Finally, call the URI. inventory :

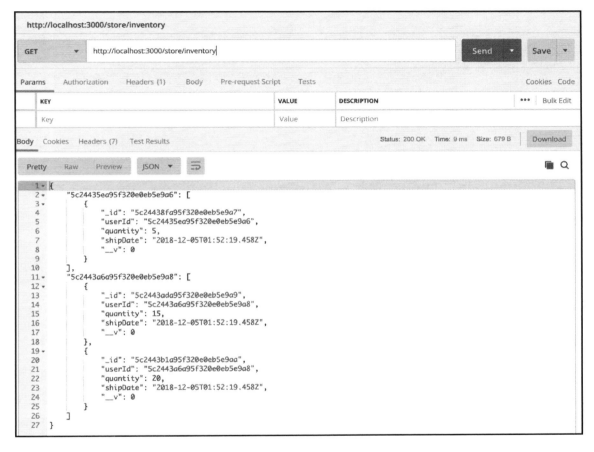

Orders inventory

6. To make sure the documents are stored on MongoDB, check them using the Robomongo client. Users will be shown the following:

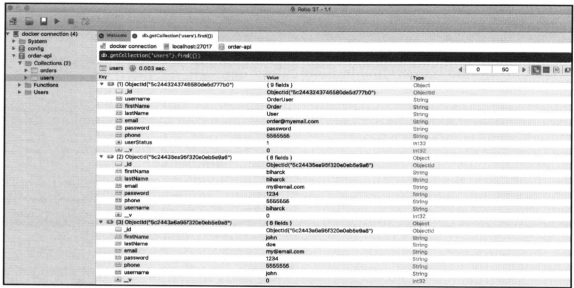

Users created and stored on MongoDB

The `orders` will be shown as follows:

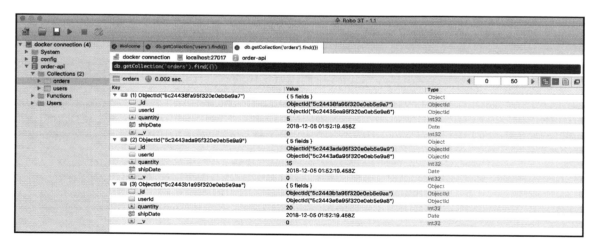

Orders created with the new users and stored on MongoDB

As we can see from all the `users` and `orders` created, our test is successful.

Summary

This chapter showed you how to change the in-memory strategy to a document-based database called MongoDB. We were able to see that we could still follow the testing strategy using the integration tests we have been evolving throughout this book, and getting fast feedback as well as writing better code.

We learned also the basics of Docker, so that we could use MongoDB through Docker rather than locally. The `order-api` is becoming more interesting and more complex. So, you are now able to create complex APIs using TypeScript and store the information on databases, such as MongoDB on Docker.

In the next chapter, we will improve our application even more by adding security to it.

Questions

1. What does ODM mean?
2. What is the difference between a traditional relational database and a document database?
3. What is `mongoose`?
4. What does a schema on mongoose mean?
5. What is the `_id` property?
6. What is Docker?
7. What is the command to spin up a docker with `mongo`?
8. What is the `mongoose` method of persisting data?
9. What is the `mongoose` method of retrieving a resource using an ID as a filter?

Further reading

To improve your knowledge with formatting strategies, check out the following books for more information:

- *JavaScript Unit Testing* (https://www.packtpub.com/web-development/javascript-unit-testing)
- *Learning Behavior-driven Development with JavaScript* (https://www.packtpub.com/application-development/learning-behavior-driven-development-javascript)
- *MongoDB 4 Quick Start Guide* (https://www.packtpub.com/big-data-and-business-intelligence/mongodb-4-quick-start-guide)
- *Node.js Web Development – Fourth Edition* (https://www.packtpub.com/web-development/nodejs-web-development-fourth-edition)
- *Web Development with MongoDB and Node – Third Edition* (https://www.packtpub.com/web-development/web-development-mongodb-and-node-third-edition)
- *MongoDB Administrator's Guide* (https://www.packtpub.com/big-data-and-business-intelligence/mongodb-administrators-guide)

9
Securing Your API

Every API needs some form of security for validating its access, request, and output options. In this chapter, we will discuss some of the authorization techniques for authenticating users by using JWT-based tokens and basic authentication. Moving on, we will use tools such as Passport and look at security best practices. This chapter will describe the importance of serving APIs with SSL, as well as how to validate data so that we don't expose sensitive information.

The following topics will be covered in this chapter:

- Authorization techniques
- Authenticating requests
- Securing API
- Validation

Technical requirements

All of the information that's required to run the code in this chapter can be found in the relevant sections. The only requirement is that you have applications such as Node.js, VS Code, TypeScript, and so on installed on your system, which we covered in Chapter 4, *Setting Up Your Development Environment*.

All of the code that's used in this chapter is available at https://github.com/ PacktPublishing/Hands-On-RESTful-Web-Services-with-TypeScript-3/tree/master/ Chapter09.

Security overview

When it comes to the subject of security, a lot of developers get confused about what kind of information should be protected and when it should be protected. Like almost everything in computer science, the answer is, it depends.

This trade-off is a crucial decision when you're developing APIs that could define either the success or failure of your API.

This chapter will show you some techniques that might help you with API security. Of course, if you want to learn even more about security, take a look at the *Further reading* section.

Using HTTPS over HTTP

In a tweet, the first and most important thing that you should do is always use HTTPS over HTTP. We know that this is sometimes difficult during the development process, because there is no valid certificate, but this should not prevent you from doing so. You can run HTTPS locally with a self-signed certificate.

For local testing, we will use OpenSSL to generate a key and the certificates for HTTPS configuration.

 The OpenSSL documentation is available at https://www.openssl.org/ docs/.

Once you have OpenSSL configured, go to the root folder of the `order-api` project and create a new certificate there. The command to create a certificate on macOS is as follows:

```
$ openssl req -newkey rsa:2048 -nodes -keyout keytemp.pem -x509 -days 365 -
out cert.pem

$ openssl rsa -in keytemp.pem -out key.pem
```

This command will generate a new key and `cert` file for you. Right after creating those files, make sure to add them as part of the `.gitignore` file if you are using **Git**.

Right after creating these files, move them to a new folder called `config`, under the root level of the `order-api` project, and change the `src/server.ts` file to allow HTTPS connections:

```
import * as fs from 'fs'
import * as https from 'https'
import app from './app'

const PORT = process.env.PORT

const httpsOptions = {
  key: fs.readFileSync('./config/key.pem'),
  cert: fs.readFileSync('./config/cert.pem'),
}
https.createServer(httpsOptions, app).listen(PORT)
```

Now, if you start the application, the URIs should be available only on the `HTTPS` level:

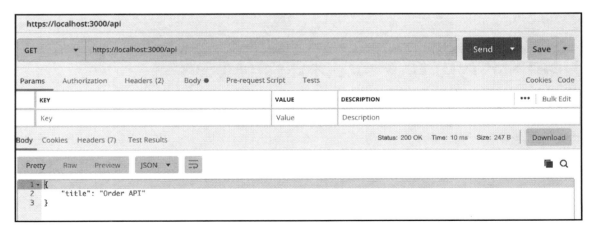

Requests over HTTPS

If we try to call HTTP, we should not be able to get the connection with the server, as follows:

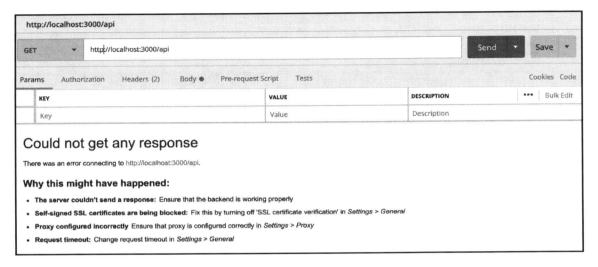

Request denied through HTTP

Improving API security

Considering that requests are now made via HTTPS, it is time to improve our security with tokens and encryption, for when we store sensitive data on the database.

Using bcrypt to encrypt a password

The next step is to make sure that the user's password will be encrypted when we store it on MongoDB. This process is really easy and only uses a new library, called bcrypt.

bcrypt is available at https://www.npmjs.com/package/bcrypt.

To use bcrypt, install it through npm:

```
$ npm install --save bcrypt
```

You also need to install the types:

```
$ npm install --save @types/bcrypt
```

Once you have installed bcrypt, as well as its types, change the 02_user.spec.ts file to create a new user on the before step, with the password encrypted:

```
...

import * as bcrypt from 'bcrypt'

...

before(async () => {
    expect(UserModel.modelName).to.be.equal('User')
    UserModel.collection.drop()
    const newUser = new UserModel(user)

    newUser.password = bcrypt.hashSync(newUser.password, 10)

    await newUser.save((error, userCreated) => {
      user._id = userCreated._id
    })
  })
```

newUser.password = bcrypt.hashSync(newUser.password, 10) will encrypt the password before storing it in the database. After that, change the addUser method on the user controller so that the password is encrypted before it is saved:

```
...

import * as bcrypt from 'bcrypt'

...


export let addUser = (req: Request, res: Response, next: NextFunction) => {
  const newUser = new UserModel(req.body)

  newUser.password = bcrypt.hashSync(newUser.password, 10)

  newUser.save((error, user) => {
    user = halson(user.toJSON()).addLink('self', `/users/${user._id}`)
    return formatOutput(res, user, 201, 'user')
  })
}
```

You may have noticed that, on MongoDB, the password is now encrypted, as shown in the following screenshot:

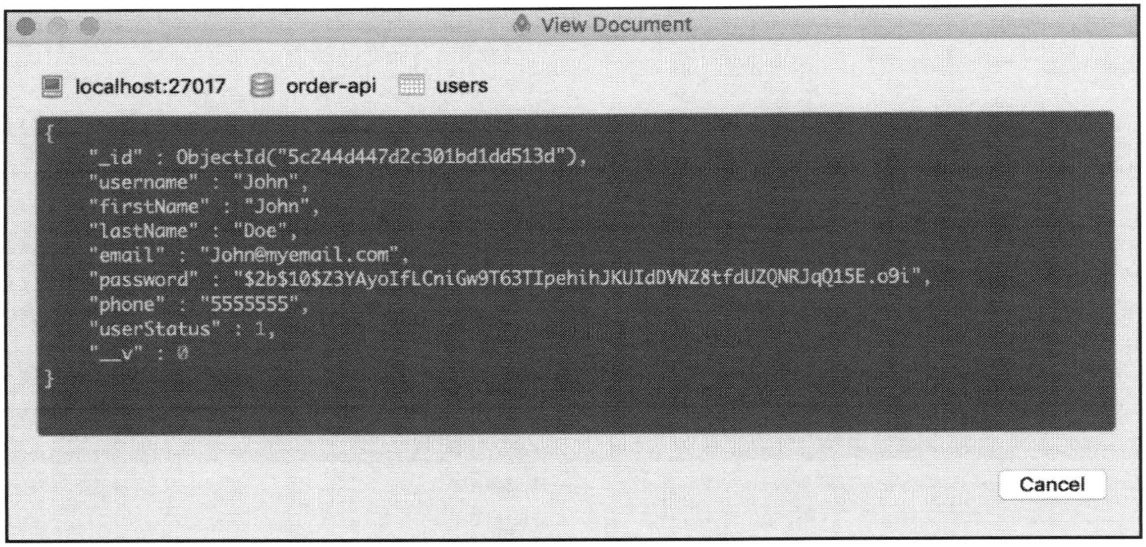

```
localhost:27017    order-api    users

{
    "_id" : ObjectId("5c244d447d2c301bd1dd513d"),
    "username" : "John",
    "firstName" : "John",
    "lastName" : "Doe",
    "email" : "John@myemail.com",
    "password" : "$2b$10$Z3YAyoIfLCniGw9T63TIpehihJKUIdDVNZ8tfdUZQNRJqQ15E.o9i",
    "phone" : "5555555",
    "userStatus" : 1,
    "__v" : 0
}
```

Password encrypted on MongoDB

Adding authentication

To get authentication on `order-api`, we will use Passport. Passport is a piece of authentication middleware for Node.js, and is considered one of the most widely used pieces of middleware for this purpose. It is known as flexible and modular. There are a lot of strategies that support authentication using a username and password, Facebook, Twitter, **JSON Web Tokens (JWTs)**, and other methods. For the purpose of this book, we will use JWT.

The first step is to install Passport and JWT:

```
$ npm install --save passport passport-jwt jsonwebtoken passport-local
```

Then, we need to install the types:

```
$ npm install --save-dev @types/jsonwebtoken @types/passport-jwt
@types/passport @types/passport-local
```

Now that we have all of the dependencies installed, we can add our login route and controller. First, let's change our tests so that they consider that we have an authentication strategy using JWT. Our `02_user.spec.js` file should look as follows:

```
'use strict'
import * as bcrypt from 'bcrypt'
import * as chai from 'chai'
import chaiHttp = require('chai-http')
import 'mocha'
import app from '../../src/app'
import { UserModel } from '../../src/schemas/user'
import { OrderAPILogger } from '../../src/utility/logger'

chai.use(chaiHttp)

const expect = chai.expect

describe('userRoute', async () => {
  const user = {
    _id: null,
    username: 'John',
    firstName: 'John',
    lastName: 'Doe',
    email: 'John@memail.com',
    password: 'password',
    phone: '5555555',
    userStatus: 1,
  }

  let token

  before((done) => {
    expect(UserModel.modelName).to.be.equal('User')

    UserModel.db.db.dropCollection('users', async (err, result) => {
      const newUser = new UserModel(user)
      newUser.password = bcrypt.hashSync(newUser.password, 10)
      newUser.save(async (error, userCreated) => {
        // tslint:disable-next-line:no-console
        console.log('criou')
        user._id = userCreated._id
        done()
      })
    })
  })

  it('should be able to login', () => {
    OrderAPILogger.logger.info('getting the login')
```

```
      return chai
        .request(app)
.get(`/users/login?username=${user.username}&password=${user.password}`)
        .then(res => {
          expect(res.status).to.be.equal(200)
          token = res.body.token
        })
  })

  it('should respond with HTTP 404 status because there is no user', () =>
{
      return chai
        .request(app)
        .get(`/users/NO_USER`)
        .set('Authorization', `Bearer ${token}`)
        .then(res => {
          expect(res.status).to.be.equal(404)
        })
  })
  it('should create a new user and retrieve it back', () => {
      user.email = 'unique_email@email.com'
      return chai
        .request(app)
        .post('/users')
        .set('Authorization', `Bearer ${token}`)
        .send(user)
        .then(res => {
          expect(res.status).to.be.equal(201)
          expect(res.body.username).to.be.equal(user.username)
        })
  })
  it('should return the user created on the step before', () => {
      return chai
        .request(app)
        .get(`/users/${user.username}`)
        .set('Authorization', `Bearer ${token}`)
        .then(res => {
          expect(res.status).to.be.equal(200)
          expect(res.body.username).to.be.equal(user.username)
        })
  })
  it('should updated the user John', () => {
      user.username = 'John_Updated'
      user.firstName = 'John Updated'
      user.lastName = 'Doe Updated'
      user.email = 'John@myemail_updated.com'
      user.password = 'password Updated'
      user.phone = '3333333'
```

```
      user.userStatus = 12

      return chai
        .request(app)
        .patch(`/users/John`)
        .set('Authorization', `Bearer ${token}`)
        .send(user)
        .then(res => {
          expect(res.status).to.be.equal(204)
        })
    })
  it('should return the user updated on the step before', () => {
    return chai
      .request(app)
      .get(`/users/${user.username}`)
      .set('Authorization', `Bearer ${token}`)
      .then(res => {
        expect(res.status).to.be.equal(200)
        expect(res.body.username).to.be.equal(user.username)
        expect(res.body.firstName).to.be.equal(user.firstName)
        expect(res.body.lastName).to.be.equal(user.lastName)
        expect(res.body.email).to.be.equal(user.email)
        expect(res.body.password).to.be.equal(user.password)
        expect(res.body.phone).to.be.equal(user.phone)
        expect(res.body.userStatus).to.be.equal(user.userStatus)
      })
  })
  it('should return 404 because the user does not exist', () => {
    user.firstName = 'Mary Jane'

    return chai
      .request(app)
      .patch(`/users/Mary`)
      .set('Authorization', `Bearer ${token}`)
      .send(user)
      .then(res => {
        expect(res.status).to.be.equal(404)
      })
  })
  it('should remove an existent user', () => {
    return chai
      .request(app)
      .del(`/users/${user.username}`)
      .set('Authorization', `Bearer ${token}`)
      .then(res => {
        expect(res.status).to.be.equal(204)
      })
  })
```

```
it('should return 404 when it is trying to remove an user because the
user does not exist', () => {
   return chai
     .request(app)
     .del(`/users/Mary`)
     .set('Authorization', `Bearer ${token}`)
     .then(res => {
       expect(res.status).to.be.equal(404)
     })
  })
})
```

Let's explain each one:

- There is a new method called `logging`, which will authenticate the user and give them back a token, so we'd like to test whether the token is being sent back to the requester:

```
it('should be able to login', () => {
  return chai
    .request(app)
.get(`/users/login?username=${user.username}&password=${user.passwo
rd}`)
    .then(res => {
      expect(res.status).to.be.equal(200)
      token = res.body.token
    })
})
```

- The following tests use a new property called `set('Authorization', `Bearer ${token}`)`, passing the JWT token to the authorization side to execute the operation on the server side:

```
it('should respond with HTTP 404 status because there is no
user', () => {
   user.email = 'unique_email@email.com'
   return chai
     .request(app)
     .get(`/users/NO_USER`)
     .set('Authorization', `Bearer ${token}`)
     .then(res => {
       expect(res.status).to.be.equal(404)
     })
})
```

The same strategy applies when we need to order URIs. The test file should look as follows:

```
'use strict'

import * as chai from 'chai'
import chaiHttp = require('chai-http')
import 'mocha'
import app from '../../src/app'
import { OrderModel } from '../../src/schemas/order'

chai.use(chaiHttp)

const expect = chai.expect

describe('orderRoute', () => {
  const order = {
    userId: 20,
    quantity: 1,
    shipDate: new Date(),
    status: 'PLACED',
    complete: false,
  }

  let orderIdCreated
  let token

  before(async () => {
    expect(OrderModel.modelName).to.be.equal('Order')
    OrderModel.collection.drop()
  })

  it('should be able to login and get the token to be used on orders
requests', async () => {
    return chai
      .request(app)
      .get('/users/login?username=John&password=password')
      .then(res => {
        expect(res.status).to.be.equal(200)
        token = res.body.token
      })
  })

  it('should respond with HTTP 404 status because there is no order', async
() => {
    return chai
      .request(app)
      .get(`/store/orders/000`)
      .set('Authorization', `Bearer ${token}`)
```

```
        .then(res => {
          expect(res.status).to.be.equal(404)
        })
  })

  it('should create a new user for Order tests and retrieve it back', async
() => {
    const user = {
      username: 'OrderUser',
      firstName: 'Order',
      lastName: 'User',
      email: 'order@myemail.com',
      password: 'password',
      phone: '5555555',
      userStatus: 1,
    }
    return chai
      .request(app)
      .post('/users')
      .set('Authorization', `Bearer ${token}`)
      .send(user)
      .then(res => {
        expect(res.status).to.be.equal(201)
        expect(res.body.username).to.be.equal(user.username)
        order.userId = res.body._id
      })
  })

  it('should create a new order and retrieve it back', async () => {
    return chai
      .request(app)
      .post(`/store/orders`)
      .set('Authorization', `Bearer ${token}`)
      .send(order)
      .then(res => {
        expect(res.status).to.be.equal(201)
        expect(res.body.userId).to.be.equal(order.userId)
        expect(res.body.complete).to.be.equal(false)
        orderIdCreated = res.body._id
      })
  })
  it('should return the order created on the step before', async () => {
    return chai
      .request(app)
      .get(`/store/orders/${orderIdCreated}`)
      .set('Authorization', `Bearer ${token}`)
      .then(res => {
        expect(res.status).to.be.equal(200)
```

```
        expect(res.body._id).to.be.equal(orderIdCreated)
        expect(res.body.status).to.be.equal(order.status)
      })
  })
  it('should return all orders so far', async () => {
    return chai
      .request(app)
      .get(`/store/orders`)
      .set('Authorization', `Bearer ${token}`)
      .then(res => {
        expect(res.status).to.be.equal(200)
        expect(res.body.length).to.be.equal(1)
      })
  })
  it('should not return orders because offset is higher than the size of
the orders array', async () => {
    return chai
      .request(app)
      .get(`/store/orders?offset=2&limit=2`)
      .set('Authorization', `Bearer ${token}`)
      .then(res => {
        expect(res.status).to.be.equal(200)
        expect(res.body.length).to.be.equal(0)
      })
  })
  it('should return the inventory for all users', async () => {
    return chai
      .request(app)
      .get(`/store/inventory?status=PLACED`)
      .set('Authorization', `Bearer ${token}`)
      .then(res => {
        expect(res.status).to.be.equal(200)
        expect(res.body[order.userId].length).to.be.equal(1)
      })
  })
  it('should remove an existing order', async () => {
    return chai
      .request(app)
      .del(`/store/orders/${orderIdCreated}`)
      .set('Authorization', `Bearer ${token}`)
      .then(res => {
        expect(res.status).to.be.equal(204)
      })
  })
  it('should return 404 when it is trying to remove an order because the
order does not exist', async () => {
    return chai
      .request(app)
```

```
         .del(`/store/orders/${orderIdCreated}`)
         .set('Authorization', `Bearer ${token}`)
         .then(res => {
           expect(res.status).to.be.equal(404)
         })
     })
   })
```

Of course, like we did in the previous chapter, if we run these tests, they will fail. Here, this is because we still don't have either the login route or the controller.

Let's fix this by creating the route on the `user.ts` route file:

```
app.route('/users/login').get(userController.login)
```

Now, we need to create the `login` method on the user controller file:

```
...
import * as jwt from 'jsonwebtoken'
...

export let login = (req: Request, res: Response, next: NextFunction) => {
  const username = req.query.username
  const password = req.query.password

  UserModel.findOne({ username: username }, (err, user) => {
    if (!user) {
      return res.status(404).send()
    }

    const validate = bcrypt.compareSync(password, user.password.valueOf())

    if (validate) {
      const body = { _id: user._id, email: user.email }

      const token = jwt.sign({ user: body }, 'top_secret')

      return res.json({ token: token })
    } else {
      return res.status(401).send()
    }
  })
}
```

We now have a method called `login` that searches for users based on their username and then compares the password that's provided with what's stored on the database using `bcrypt`. If everything is fine, we will create a new JWT and can send this token back to the requester. JWT is a JSON-based open standard for creating access tokens that assert a number of claims.

The next step is to create a process that guarantees that if you do not have a valid token, you won't be able to access private routes.

In the `utility` folder, create a new file called `passportConfiguration.ts` with the following content:

```
import * as passport from 'passport'
import { Strategy } from 'passport-jwt'
import { ExtractJwt } from 'passport-jwt'

export class PassportConfiguration {
  constructor() {
    passport.use(
      new Strategy(
        {
          secretOrKey: 'top_secret',
          jwtFromRequest: ExtractJwt.fromAuthHeaderAsBearerToken(),
        },
        async (token, done) => {
          try {
            return done(null, token.user)
          } catch (error) {
            done(error)
          }
        }
      )
    )
  }
}
```

This class will check whether the JWT is valid or not with the `top_secret` secret.

Note that in real-word services, the secret must be more complex and should be stored in a vault or using any other strategy than being hard-coded, like in this example.

Now, let's include this validation on our routes. First up is `src/routes/user.ts`:

```
import * as passport from 'passport'
import * as userController from '../controllers/user'
import { PassportConfiguration } from '../utility/passportConfiguration'

export class UserRoute extends PassportConfiguration {
  public routes(app): void {
    app
      .route('/users')
      .post(
        passport.authenticate('jwt', { session: false }),
        userController.addUser
      )
    app.route('/users/login').get(userController.login)
    app
      .route('/users/:username')
      .patch(
        passport.authenticate('jwt', { session: false }),
        userController.updateUser
      )
    app
      .route('/users/:username')
      .delete(
        passport.authenticate('jwt', { session: false }),
        userController.removeUser
      )
    app
      .route('/users/:username')
      .get(
        passport.authenticate('jwt', { session: false }),
        userController.getUser
      )
  }
}
```

Then, we will do the same for `src/routes/order.ts`:

```
import * as passport from 'passport'
import * as orderController from '../controllers/order'
import { PassportConfiguration } from '../utility/passportConfiguration'

export class OrderRoute extends PassportConfiguration {
  public routes(app): void {
    app
      .route('/store/inventory')
      .get(
        passport.authenticate('jwt', { session: false }),
```

```
      orderController.getInventory
    )
  app
    .route('/store/orders')
    .post(
      passport.authenticate('jwt', { session: false }),
      orderController.addOrder
    )
  app
    .route('/store/orders')
    .get(
      passport.authenticate('jwt', { session: false }),
      orderController.getAllOrders
    )
  app
    .route('/store/orders/:id')
    .get(
      passport.authenticate('jwt', { session: false }),
      orderController.getOrder
    )
  app
    .route('/store/orders/:id')
    .delete(
      passport.authenticate('jwt', { session: false }),
      orderController.removeOrder
    )
  }
}
```

Now, every request is going to pass to the authentication method to check and validate whether the requester is authorized to complete the action or not.

Note that only /api and /login are not controlled by JWT, because we want them to be like this.

Now, run the tests again:

```
$ npm run test

baseRoute
    should respond with HTTP 200 status (50ms)
    should respond with success message

  userRoute
    should be able to login (101ms)
    should respond with HTTP 404 status because there is no user
    should create a new user and retrieve it back (109ms)
    should return the user created on the step before
```

```
should updated the user John
should return the user updated on the step before
should return 404 because the user does not exist
should remove an existent user
should return 404 when it is trying to remove an user because the user
does not exist

userRoute
should be able to login and get the token to be used on orders requests
(85ms)
should respond with HTTP 404 status because there is no order
should create a new user for Order tests and retrieve it back (87ms)
should create a new order and retrieve it back (40ms)
should return the order created on the step before
should return all orders so far
should not return orders because offset is higher than the size of the
orders array
should return the inventory for all users
should remove an existing order
should return 404 when it is trying to remove an order because the
order does not exist


21 passing (746ms)
```

Email validation

Now that we have the login action, we should make sure that no more than one user is allowed to use the same email address.

There is a library we could use to validate it called **mongoose-validator**.

To use it, install it using npm:

```
$ npm install --save mongoose-unique-validator
```

Then, install the necessary types:

```
$ npm install --save @types/mongoose-unique-validator
```

Using it is simple – just add the validation to the property on the schema:

```
import { Document, Model, model, Schema } from 'mongoose'
import * as uniqueValidator from 'mongoose-unique-validator'
import { default as User } from '../models/user'

export interface UserModel extends User, Document {}

export const UserSchema: Schema = new Schema({
  firstName: String,
  lastName: String,
  email: { type: String, unique: true },
  password: String,
  phone: String,
  userStatus: Number,
  username: String,
})

UserSchema.plugin(uniqueValidator)

export const UserModel: Model<UserModel> = model<UserModel>('User',
UserSchema)
```

Then, add the user `saves` operation:

```
export let addUser = (req: Request, res: Response, next: NextFunction) => {
  const newUser = new UserModel(req.body)

  newUser.password = bcrypt.hashSync(newUser.password, 10)

  newUser.save((error, user) => {
    if (error) {
      return res.status(500).send(error)
    }
    user = halson(user.toJSON()).addLink('self', `/users/${user._id}`)
    return formatOutput(res, user, 201, 'user')
  })
}
```

Now, if someone tries to sign up with an email address that is already in use, we will get the following output:

Error when there is already someone with this email address

Manual testing

Now, it is time to test security with Postman. Start the application and try to create a new user in the same way as before:

Unauthorized when trying to create a new user without authentication

Since you are not authenticated, you are not allowed to create a new user. Go to the Robomongo application and insert a new user manually with the following content:

```
{
    "username" : "John",
    "firstName" : "John",
    "lastName" : "Doe",
    "email" : "John@myemail.com",
    "password" :
"$2b$10$Z3YAyoIfLCniGw9T63TIpehihJKUIdDVNZ8tfdUZQNRJqQ15E.o9i",
    "phone" : "5555555",
    "userStatus" : 1
}
```

Now, go back to Postman and call the login, as follows:

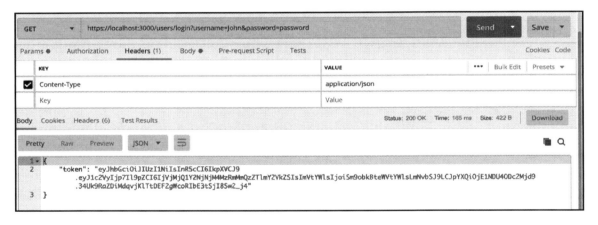

Getting a JWT token from the login

The output from the login operation is a JWT token. You can validate and see the content on https://jwt.io/. Go to the debugger section, paste the token you got back from the login, and add the secret called `top_secret` on `your-256-bit-secret`. You will then see the JWT content there, as follows:

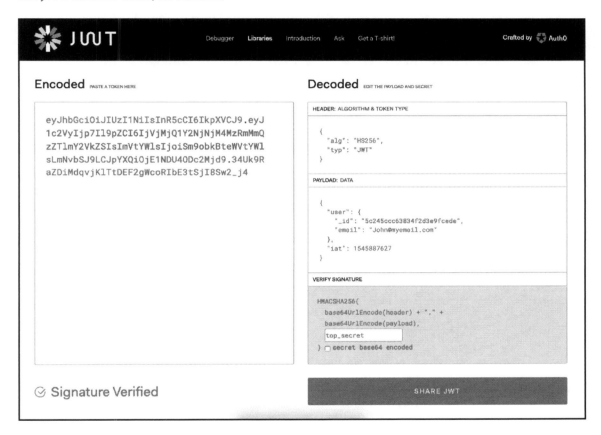

Validating JWT on the jwt.io website

Now, you can go back to Postman, add this token on the **HEADER**, and call `create user` again:

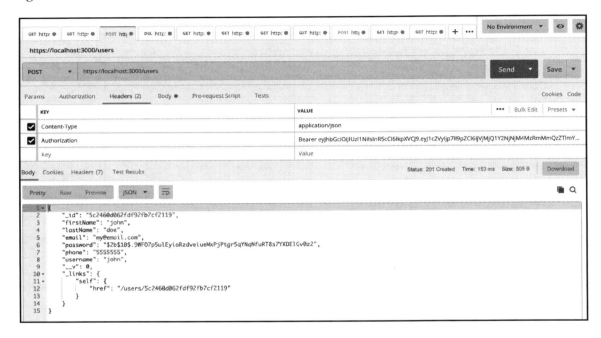

Getting access to create a user with a JWT

Now, you are able to perform this action and the other ones that are marked as private with the JWT. Note that the `Bearer` tag has to be included on the authorization header as part of the token value.

Summary

Even though this is just a small part of the huge world of security, you have gained an understanding of some of the best practices when it comes to developing APIs, such as authentication, authorization, HTTPS, and so on.

In this chapter, we covered a simple way to encrypt sensitive information that is persisted outside the application and perhaps over the internet.

As part of the authentication process, you were able to use Passport for JWT and validate whether the user has a valid token or not.

In the next chapter, will show you how to improve logging and error handling.

Questions

1. What do we mean by authentication?
2. What do we mean by authorization?
3. What is JWT?
4. What is `bcrypt`?
5. What is `mongoose-unique-validator`?
6. Why should you use HTTPS instead of HTTP?
7. Is the code base the best place to store the JWT secret?

Further reading

To improve your knowledge regarding what was covered in this chapter, the following books are recommended, as they will be helpful for the upcoming chapters:

- *Cybersecurity - Attack and Defense Strategies* (https://www.packtpub.com/networking-and-servers/cybersecurity-attack-and-defense-strategies)
- *Hands-On Cybersecurity for Architects* (https://www.packtpub.com/networking-and-servers/hands-cybersecurity-architects)

10
Error Handling and Logging

Debugging can become a burden if you do not have meaningful error messages or if you do not handle errors in the correct way. In this chapter, we are going to focus on how to handle errors, starting with how to catch them and how to describe what an error is and is not. Without meaningful error messages, catching an error is usually hard to debug. Note that error messages should only describe the error and should not expose any sensitive data inside the error. We are going to talk about how to write understandable error messages with minimal information. At the end of the chapter, we will explore logging mechanisms in order to track what is going on in our web services, use them for attack detection, and to improve our code base in terms of quality and validity.

The following topics will be covered in this chapter:

- Error handling
- Error messages
- Logging

Technical requirements

All the information required to run the code from this chapter is provided by the chapter itself. The only necessity is to get the previous installation process from the previous chapter, that is Node.js, VS Code, and TypeScript.

The code base is available at: `https://github.com/PacktPublishing/Hands-On-RESTful-Web-Services-with-TypeScript-3/tree/master/Chapter10`.

Error handling

Error handling is the process of an application catching errors and processing them. These errors usually occur during runtime of an application. It does not matter if an error is synchronous or asynchronous. A default error handler is embedded by default in Express.js hence, it's not required for you to write a specific error handler to get started on a new application as we did.

It is necessary for Express.js to catch any errors that occur while running route handlers and middleware, otherwise errors will be lost. As mentioned, there are some default error handlers for errors that occur in synchronous code inside route handlers and middleware. In the case of default error handlers, no extra work is needed. In other words, if asynchronous code throws an error, then Express.js will catch it and process it by default, as in the following example:

```
export let getSomething = (req: Request, res: Response, next: NextFunction)
=> {
  throw new Error("doh! Something went wrong :( ")
})
```

So, if you have asynchronous functions that are invoked by route handlers and middleware, they must be passed to the `next()` function. That way, Express.js will catch and process them automatically, as in the following example:

```
...
import * as fs from 'fs'
...

export let getSomething = (req: Request, res: Response, next: NextFunction)
=> {
  fs.readFile("/PATH_TO_A_FILE_THAT_NOT_EXISTS", (error, data) => {
    if (error) {
      // no worries, just pass the errors to Express through next function
      next(err);
    }
    else {
      res.status(200).send(data)
    }
  })
})
```

There is also another possibility available, which is evolving the code in a try...catch block in order to catch errors in asynchronous code and pass them back to Express.js:

 If a try...catch block is ignored, Express.js will not be able to catch any errors because it is no longer part of the asynchronous handler scope.

```
export let getApi = (req: Request, res: Response, next: NextFunction) => {
  Promise.resolve()
    .then(() => {
      throw new Error('doh! Something went wrong :( ')
    })
    .catch(next)
}
```

For instance, in the preceding code, the error will be passed to Express.js on the catch scope.

Writing custom error handlers

To write error handlers using Express.js, you can follow almost the same template as a middleware function, for example:

```
this.app.use((err, req, res, next) => {
  console.error(err.stack)
  res.status(500).send('My custom error handler with 500 error...')
})
```

 The error handler must be the last app definition after routes and other calls.

Also, you can define as many error handlers as you want, for instance, one for logging, another one for client errors, and another one for a catch-all error, as follows:

```
this.app.use(errorHandler.logging)
this.app.use(errorHandler.clientErrorHandler)
this.app.use(errorHandler.errorHandler)
```

For logging purposes, we want to catch the log and pass the error on the `next()` function, as follows:

```
export let logging = (err: Error, req: Request, res: Response, next:
NextFunction) => {
  logger.error(err)
  next(err)
}
```

On client error, if the error is of the `xhr` type, we will send a `500` error to the requester, and if it is not, we will again pass the error on the `next()` function, as follows:

```
export let clientError = (err: Error, req: Request, res: Response, next:
NextFunction) => {
  if (req.xhr) {
    res.status(500).send({ error: err.message })
  } else {
    next(err)
  }
}
```

Finally, if none of the previous steps catch the error, the default error handler will catch everything and send a `500` error, as follows:

```
export let errorHandler = (err: Error, req: Request, res: Response, next:
NextFunction) => {
  res.status(500).send({ error: err.message })
}
```

Adding an error handler to order-api

Since we now know how to implement an error handler, we can apply the error handler strategy to our `order-api` project. First, create a file called `errorHandler.ts` in the `src/utility` folder, with the following content:

```
import { NextFunction, Request, Response } from 'express'

export let logging = (
  err: Error,
  req: Request,
  res: Response,
  next: NextFunction
) => {
  console.log(err)
  next(err)
}
```

```
export let clientErrorHandler = (
  err: Error,
  req: Request,
  res: Response,
  next: NextFunction
) => {
  if (req.xhr) {
    res.status(500).send({ error: 'Something failed!' })
  } else {
    next(err)
  }
}

export let errorHandler = (
  err: Error,
  req: Request,
  res: Response,
  next: NextFunction
) => {
  res.status(500).send({ error: err.message })
}
```

As you can see, we used the same idea of logging, client error, and default error handler code, which will be used throughout the code base.

Now, change the `app.ts` file to tell Express.js which error handler functions should be used, as follows:

```
import * as bodyParser from 'body-parser'
import * as express from 'express'
import * as mongoose from 'mongoose'
import { APIRoute } from '../src/routes/api'
import { OrderRoute } from '../src/routes/order'
import { UserRoute } from '../src/routes/user'
import * as errorHandler from '../src/utility/errorHandler'

class App {
  public app: express.Application
  public userRoutes: UserRoute = new UserRoute()
  public apiRoutes: APIRoute = new APIRoute()
  public orderRoutes: OrderRoute = new OrderRoute()
  public mongoUrl: string = 'mongodb://localhost/order-api'

  constructor() {
    this.app = express()
    this.app.use(bodyParser.json())
    this.userRoutes.routes(this.app)
    this.apiRoutes.routes(this.app)
```

```
      this.orderRoutes.routes(this.app)
      this.mongoSetup()
      this.app.use(errorHandler.logging)
      this.app.use(errorHandler.clientErrorHandler)
      this.app.use(errorHandler.errorHandler)
    }

    private mongoSetup(): void {
      mongoose.connect(
        this.mongoUrl,
        { useNewUrlParser: true }
      )
    }
  }

  export default new App().app
```

Remember to include the error handler functions at the end of the `app.use` definition.

We now have a custom error handler strategy. Every time that something goes wrong, first the logger handler will catch the error, then it will pass it to the client error handler, and if the error is an `xhr` type error, it will send a `500` error; otherwise, it will pass it to the catch-all error message.

Error messages

We have a well-defined error handler, as we saw in the previous section, so we can now write meaningful error messages that will help you with troubleshooting and make sure that sensitive information is not exposed.

Error messages will depend on the type of application you are coding for. As an example, think about what kind of error message you would like to see to help you solve an issue or understand what is happening.

We will apply some ideas and you can then go further and improve the error messages for the `order-api` application.

First, we will change the `order.ts` controller file, as follows:

```
import { NextFunction, Request, Response } from 'express'
import * as halson from 'halson'
import * as _ from 'lodash'
import { OrderModel } from '../schemas/order'
import { UserModel } from '../schemas/User'
```

```
import { formatOutput } from '../utility/orderApiUtility'

export let getOrder = (req: Request, res: Response, next: NextFunction) =>
{
  const id = req.params.id

  OrderModel.findById(id, (err, order) => {
    if (!order) {
      return next(new Error(`Order ${id} not found.`))
    }
    order = halson(order.toJSON()).addLink('self',
`/store/orders/${order.id}`)
    return formatOutput(res, order, 200, 'order')
  })
}

export let getAllOrders = (req: Request, res: Response, next: NextFunction)
=> {
  const limit = Number(req.query.limit) || 0
  const offset = Number(req.query.offset) || 0

  OrderModel.find({}, null, { skip: offset, limit: limit }).then(orders =>
{
    if (orders) {
      orders = orders.map(order => {
        return halson(order.toJSON())
          .addLink('self', `/store/orders/${order.id}`)
          .addLink('user', {
            href: `/users/${order.userId}`,
          })
      })
    }
    return formatOutput(res, orders, 200, 'order')
  })
}

export let addOrder = (req: Request, res: Response, next: NextFunction) =>
{
  const userId = req.body.userId

  UserModel.findById(userId, (err, user) => {
    if (!user) {
      throw new Error(`There is no user with the userId ${userId}`)
    }

    const newOrder = new OrderModel(req.body)

    newOrder.save((error, order) => {
```

```
      order = halson(order.toJSON())
        .addLink('self', `/store/orders/${order._id}`)
        .addLink('user', {
          href: `/users/${order.userId}`,
        })

      return formatOutput(res, order, 201, 'order')
    })
  })
}

export let removeOrder = (req: Request, res: Response, next: NextFunction)
=> {
  const id = req.params.id

  OrderModel.findById(id, (err, order) => {
    if (!order) {
      return res.status(404).send()
    }
    order.remove(error => {
      res.status(204).send()
    })
  })
}

export let getInventory = (req: Request, res: Response, next: NextFunction)
=> {
  const status = req.query.status

  OrderModel.find({ status: status }, (err, orders) => {
    orders = _.groupBy(orders, 'userId')
    return formatOutput(res, orders, 200, 'inventory')
  })
}
```

As you can see, we decided not to add a lot of text on the error handler for order controller. The reason for this is that, sometimes, less information is better than a lot of useless words.

The same idea was applied to the user controller in the following code snippet:

```
import * as bcrypt from 'bcrypt'
import { NextFunction, Request, Response } from 'express'
import * as halson from 'halson'
import * as jwt from 'jsonwebtoken'
import { UserModel } from '../schemas/User'
import { formatOutput } from '../utility/orderApiUtility'

export let getUser = (req: Request, res: Response, next: NextFunction) => {
```

```
  const username = req.params.username

  UserModel.findOne({ username: username }, (err, user) => {
    if (!user) {
      return res.status(404).send()
    }

    user = user.toJSON()
    user._id = user._id.toString()

    user = halson(user).addLink('self', `/users/${user._id}`)
    return formatOutput(res, user, 200, 'user')
  })
}

export let addUser = (req: Request, res: Response, next: NextFunction) => {
  const newUser = new UserModel(req.body)

  newUser.password = bcrypt.hashSync(newUser.password, 10)

  newUser.save((error, user) => {
    if (error) {
      return res.status(500).send(error)
    }
    user = halson(user.toJSON()).addLink('self', `/users/${user._id}`)
    return formatOutput(res, user, 201, 'user')
  })
}

export let updateUser = (req: Request, res: Response, next: NextFunction)
=> {
  const username = req.params.username

  UserModel.findOne({ username: username }, (err, user) => {
    if (!user) {
      return res.status(404).send()
    }

    user.username = req.body.username || user.username
    user.firstName = req.body.firstName || user.firstName
    user.lastName = req.body.lastName || user.lastName
    user.email = req.body.email || user.email
    user.password = req.body.password || user.password
    user.phone = req.body.phone || user.phone
    user.userStatus = req.body.userStatus || user.userStatus

    user.save(error => {
      res.status(204).send()
```

```
      })
    })
  }

  export let removeUser = (req: Request, res: Response, next: NextFunction)
  => {
    const username = req.params.username

    UserModel.findOne({ username: username }, (err, user) => {
      if (!user) {
        return res.status(404).send()
      }

      user.remove(error => {
        res.status(204).send()
      })
    })
  }

  export let login = (req: Request, res: Response, next: NextFunction) => {
    const username = req.query.username
    const password = req.query.password

    UserModel.findOne({ username: username }, (err, user) => {
      if (!user) {
        return res.status(404).send()
      }

      const validate = bcrypt.compareSync(password, user.password.valueOf())

      if (validate) {
        const body = { _id: user._id, email: user.email }

        const token = jwt.sign({ user: body }, 'top_secret')

        return res.json({ token: token })
      } else {
        return res.status(401).send()
      }
    })
  }
```

Of course, even when adopting this error handler strategy, our tests should pass as before:

```
$ npm run test
```

This will produce the following output:

```
baseRoute
    should respond with HTTP 200 status (126ms)
    should respond with success message

  userRoute
    should be able to login (117ms)
    should respond with HTTP 404 status because there is no user
    should create a new user and retrieve it back (121ms)
    should return the user created on the step before
    should updated the user John (45ms)
    should return the user updated on the step before
    should return 404 because the user does not exist
    should remove an existent user
    should return 404 when it is trying to remove an user because the user
does not exist

  userRoute
    should be able to login and get the token to be used on orders requests
(101ms)
    should respond with HTTP 404 status because there is no order (63ms)
    should create a new user for Order tests and retrieve it back (84ms)
    should create a new order and retrieve it back
    should return the order created on the step before
    should return all orders so far
    should not return orders because offset is higher than the size of the
orders array
    should return the inventory for all users
    should remove an existing order
    should return 404 when it is trying to remove an order because the
order does not exist


  21 passing (1s)
```

Logging

A good logging strategy will also help you to troubleshoot issues in addition to a good error handler strategy. Some pieces of information don't need to be sent to the end user, but it would be helpful for you to be able to trace all errors in logging files in order to understand what is going on.

It is recommended that you log every operation of an application. This log strategy means logging more than just errors that occur; it means that critical operations should also be logged, for instance, so that you can have more control over who is doing what for audit purposes. Applications can log at a code level (debugging, for instance) and at a user request level for audits and forbidden access.

Currently, `winston` is the most widely-used library for logging with Node.js, so we will use `winston` to implement our log strategy in this section.

To install `winston`, as with other Node.js packages we have already installed, use `npm`, as follows:

```
$ npm install --save winston express-winston
```

You can also use the following:

```
$ npm install --save-dev @types/winston @types/express-winston
```

With the libraries already installed, create a file called `logger.ts` in the `src/utilit` folder with the following content:

```
import { createLogger, format, transports } from 'winston'
const { combine, timestamp, label, prettyPrint, printf } = format

export class OrderAPILogger {
  public static myFormat = printf(info => {
    return `[${info.timestamp}] [${info.level}] => ${info.message}`
  })

  public static logger = createLogger({
    level: 'info',
    format: combine(
      label({ label: 'order-api errors' }),
      timestamp(),
      OrderAPILogger.myFormat
    ),

    transports: [
      new transports.File({ filename: 'aggregated.log' }),
```

```
      new transports.Console(),
    ],
  })
}
```

As you can see, there are two different kinds of transports—one for writing the output in a file called `combined.log`, and another one to log messages on the console:

```
transports: [
  new transports.File({ filename: 'combined.log' }),
  new transports.Console(),
],
```

To use the custom `logger` class, we can import the class as follows:

```
import { OrderAPILogger } from '../utility/logger'
```

We can also use the `logger` as follows:

```
OrderAPILogger.logger.info(`my message...`)
```

We will also add a default error logger on the app level. Change the `app.ts` file, adding the logger strategy there, as follows:

```
import * as bodyParser from 'body-parser'
import * as express from 'express'
import * as expressWinston from 'express-winston'
import * as mongoose from 'mongoose'
import * as winston from 'winston'
import { APIRoute } from '../src/routes/api'
import { OrderRoute } from '../src/routes/order'
import { UserRoute } from '../src/routes/user'
import * as errorHandler from '../src/utility/errorHandler'

class App {
  public app: express.Application
  public userRoutes: UserRoute = new UserRoute()
  public apiRoutes: APIRoute = new APIRoute()
  public orderRoutes: OrderRoute = new OrderRoute()
  public mongoUrl: string = 'mongodb://localhost/order-api'

  constructor() {
    this.app = express()
    this.app.use(bodyParser.json())
    this.userRoutes.routes(this.app)
    this.apiRoutes.routes(this.app)
    this.orderRoutes.routes(this.app)
    this.mongoSetup()
```

```
    this.app.use(
      expressWinston.errorLogger({
        transports: [new winston.transports.Console()],
      })
    )
    this.app.use(errorHandler.logging)
    this.app.use(errorHandler.clientErrorHandler)
    this.app.use(errorHandler.errorHandler)
  }

  private mongoSetup(): void {
    mongoose.connect(
      this.mongoUrl,
      { useNewUrlParser: true }
    )
  }
}

export default new App().app
```

Following on from this, on the `order.ts` controller, add some logger messages, as follows:

```
import { NextFunction, Request, Response } from 'express'
import * as halson from 'halson'
import * as _ from 'lodash'
import { OrderModel } from '../schemas/order'
import { UserModel } from '../schemas/User'
import { OrderAPILogger } from '../utility/logger'
import { formatOutput } from '../utility/orderApiUtility'

export let getOrder = (req: Request, res: Response, next: NextFunction) =>
{
  const id = req.params.id

  OrderAPILogger.logger.info(`[GET] [/store/orders/] ${id}`)

  OrderModel.findById(id, (err, order) => {
    if (!order) {
      OrderAPILogger.logger.info(
        `[GET] [/store/orders/:{orderId}] Order ${id} not found.`
      )
      return next(new Error(`Order ${id} not found.`))
    }
    order = halson(order.toJSON()).addLink('self',
`/store/orders/${order.id}`)
    return formatOutput(res, order, 200, 'order')
  })
}
```

```
export let getAllOrders = (req: Request, res: Response, next: NextFunction)
=> {
  const limit = Number(req.query.limit) || 0
  const offset = Number(req.query.offset) || 0

  OrderAPILogger.logger.info(`[GET] [/store/orders/]`)

  OrderModel.find({}, null, { skip: offset, limit: limit }).then(orders =>
{
    if (orders) {
      orders = orders.map(order => {
        return halson(order.toJSON())
          .addLink('self', `/store/orders/${order.id}`)
          .addLink('user', {
            href: `/users/${order.userId}`,
          })
      })
    }
    return formatOutput(res, orders, 200, 'order')
  })
}

export let addOrder = (req: Request, res: Response, next: NextFunction) =>
{
  const userId = req.body.userId

  OrderAPILogger.logger.info(`[POST] [/store/orders/] ${userId}`)

  UserModel.findById(userId, (err, user) => {
    if (!user) {
      OrderAPILogger.logger.info(
        `[POST] [/store/orders/] There is no user with the userId
${userId}`
      )
      throw new Error(`There is no user with the userId ${userId}`)
    }

    const newOrder = new OrderModel(req.body)

    OrderAPILogger.logger.info(`[POST] [/store/orders/] ${newOrder}`)

    newOrder.save((error, order) => {
      order = halson(order.toJSON())
        .addLink('self', `/store/orders/${order._id}`)
        .addLink('user', {
          href: `/users/${order.userId}`,
        })
```

```
        return formatOutput(res, order, 201, 'order')
      })
    })
  }

  export let removeOrder = (req: Request, res: Response, next: NextFunction)
  => {
    const id = req.params.id

    OrderAPILogger.logger.warn(`[DELETE] [/store/orders/] ${id}`)

    OrderModel.findById(id, (err, order) => {
      if (!order) {
        OrderAPILogger.logger.warn(
          `[DELETE] [/store/orders/:{orderId}] Order id ${id} not found`
        )
        return res.status(404).send()
      }
      order.remove(error => {
        res.status(204).send()
      })
    })
  }

  export let getInventory = (req: Request, res: Response, next: NextFunction)
  => {
    const status = req.query.status

    OrderAPILogger.logger.info(`[GET] [/store/inventory/] ${status}`)

    OrderModel.find({ status: status }, (err, orders) => {
      orders = _.groupBy(orders, 'userId')
      return formatOutput(res, orders, 200, 'inventory')
    })
  }
```

We will also include a log in the `user.ts` controller, as follows:

```
  import * as bcrypt from 'bcrypt'
  import { NextFunction, Request, Response } from 'express'
  import * as halson from 'halson'
  import * as jwt from 'jsonwebtoken'
  import { UserModel } from '../schemas/User'
  import { OrderAPILogger } from '../utility/logger'
  import { formatOutput } from '../utility/orderApiUtility'

  export let getUser = (req: Request, res: Response, next: NextFunction) => {
    const username = req.params.username
```

```
    OrderAPILogger.logger.info(`[GET] [/users] ${username}`)

  UserModel.findOne({ username: username }, (err, user) => {
    if (!user) {
      OrderAPILogger.logger.info(
        `[GET] [/users/:{username}] user with username ${username} not
found`
      )
      return res.status(404).send()
    }

    user = user.toJSON()
    user._id = user._id.toString()

    user = halson(user).addLink('self', `/users/${user._id}`)
    return formatOutput(res, user, 200, 'user')
  })
}

export let addUser = (req: Request, res: Response, next: NextFunction) => {
  const newUser = new UserModel(req.body)

  OrderAPILogger.logger.info(`[POST] [/users] ${newUser}`)

  newUser.password = bcrypt.hashSync(newUser.password, 10)

  newUser.save((error, user) => {
    if (error) {
      OrderAPILogger.logger.info(
        `[POST] [/users] something went wrong when saving a new user ${
          newUser.username
        } | ${error.message}`
      )
      return res.status(500).send(error)
    }
    user = halson(user.toJSON()).addLink('self', `/users/${user._id}`)
    return formatOutput(res, user, 201, 'user')
  })
}

export let updateUser = (req: Request, res: Response, next: NextFunction)
=> {
  const username = req.params.username

  OrderAPILogger.logger.info(`[PATCH] [/users] ${username}`)

  UserModel.findOne({ username: username }, (err, user) => {
    if (!user) {
```

```
      OrderAPILogger.logger.info(
        `[PATCH] [/users/:{username}] user with username ${username} not
found`
      )
      return res.status(404).send()
    }

    user.username = req.body.username || user.username
    user.firstName = req.body.firstName || user.firstName
    user.lastName = req.body.lastName || user.lastName
    user.email = req.body.email || user.email
    user.password = req.body.password || user.password
    user.phone = req.body.phone || user.phone
    user.userStatus = req.body.userStatus || user.userStatus

    user.save(error => {
      res.status(204).send()
    })
  })
}

export let removeUser = (req: Request, res: Response, next: NextFunction)
=> {
  const username = req.params.username

  OrderAPILogger.logger.warn(`[DELETE] [/users] ${username}`)

  UserModel.findOne({ username: username }, (err, user) => {
    if (!user) {
      OrderAPILogger.logger.info(
        `[DELETE] [/users/:{username}] user with username ${username} not
found`
      )
      return res.status(404).send()
    }

    user.remove(error => {
      res.status(204).send()
    })
  })
}

export let login = (req: Request, res: Response, next: NextFunction) => {
  const username = req.query.username
  const password = req.query.password

  UserModel.findOne({ username: username }, (err, user) => {
    if (!user) {
```

```
      OrderAPILogger.logger.info(
        `[GET] [/users/login] nouser found with the username ${username}`
      )
      return res.status(404).send()
    }

    const validate = bcrypt.compareSync(password, user.password.valueOf())

    if (validate) {
      const body = { _id: user._id, email: user.email }

      const token = jwt.sign({ user: body }, 'top_secret')

      return res.json({ token: token })
    } else {
      OrderAPILogger.logger.info(
        `[GET] [/users/login] user not authorized ${username}`
      )
      return res.status(401).send()
    }
  })
}
```

Running the tests, you should be able to see the output either on the console or in the aggregated.log file:

```
baseRoute
(node:32071) DeprecationWarning: collection.ensureIndex is deprecated. Use
createIndexes instead.
    should respond with HTTP 200 status (114ms)
    should respond with success message

  userRoute
    should be able to login (113ms)
[2018-12-27T16:40:23.527Z] [info] => [GET] [/users] NO_USER
[2018-12-27T16:40:23.532Z] [info] => [GET] [/users/:{username}] user with
username NO_USER not found
    should respond with HTTP 404 status because there is no user
[2018-12-27T16:40:23.577Z] [info] => [POST] [/users] { _id: null,
  username: 'John',
  firstName: 'John',
  lastName: 'Doe',
  email: 'John@memail.com',
  password: 'password',
  phone: '5555555',
  userStatus: 1 }
    should create a new user and retrieve it back (133ms)
[2018-12-27T16:40:23.670Z] [info] => [GET] [/users] John
```

```
    should return the user created on the step before
[2018-12-27T16:40:23.678Z] [info] => [PATCH] [/users] John
    should updated the user John
[2018-12-27T16:40:23.699Z] [info] => [GET] [/users] John_Updated
    should return the user updated on the step before
[2018-12-27T16:40:23.704Z] [info] => [PATCH] [/users] Mary
[2018-12-27T16:40:23.709Z] [info] => [PATCH] [/users/:{username}] user with
username Mary not found
    should return 404 because the user does not exist
[2018-12-27T16:40:23.711Z] [warn] => [DELETE] [/users] John_Updated
    should remove an existent user
[2018-12-27T16:40:23.724Z] [warn] => [DELETE] [/users] Mary
[2018-12-27T16:40:23.727Z] [info] => [DELETE] [/users/:{username}] user
with username Mary not found
    should return 404 when it is trying to remove an user because the user
does not exist

  userRoute
    should be able to login and get the token to be used on orders requests
(125ms)
[2018-12-27T16:40:23.857Z] [info] => [GET] [/store/orders/] 000
[2018-12-27T16:40:23.857Z] [info] => [GET] [/store/orders/:{orderId}] Order
000 not found.
{"level":"error","message":"middlewareError","meta":{"error":{},"level":"er
ror","message":"uncaughtException: Order 000 not found.\nError: Order 000
not found.\n at order_1.OrderModel.findById
(/Users/biharck/Developer/Hands-On-RESTful-Web-Services-with-
TypeScript-3/Chapter10/order-api/src/controllers/order.ts:19:19)\n at
/Users/biharck/Developer/Hands-On-RESTful-Web-Services-with-
TypeScript-3/Chapter10/order-
api/node_modules/mongoose/lib/model.js:4645:16\n at
model.Query.Query._findOne (/Users/biharck/Developer/Hands-On-RESTful-Web-
Services-with-TypeScript-3/Chapter10/order-
api/node_modules/mongoose/lib/query.js:1961:5)\n at process.nextTick
(/Users/biharck/Developer/Hands-On-RESTful-Web-Services-with-
TypeScript-3/Chapter10/order-api/node_modules/kareem/index.js:369:33)\n at
process._tickCallback
(internal/process/next_tick.js:61:11)","stack":"Error: Order 000 not
found.\n at order_1.OrderModel.findById (/Users/biharck/Developer/Hands-On-
RESTful-Web-Services-with-TypeScript-3/Chapter10/order-
api/src/controllers/order.ts:19:19)\n at /Users/biharck/Developer/Hands-On-
RESTful-Web-Services-with-TypeScript-3/Chapter10/order-
api/node_modules/mongoose/lib/model.js:4645:16\n at
model.Query.Query._findOne (/Users/biharck/Developer/Hands-On-RESTful-Web-
Services-with-TypeScript-3/Chapter10/order-
api/node_modules/mongoose/lib/query.js:1961:5)\n at process.nextTick
(/Users/biharck/Developer/Hands-On-RESTful-Web-Services-with-
TypeScript-3/Chapter10/order-api/node_modules/kareem/index.js:369:33)\n at
```

process._tickCallback
(internal/process/next_tick.js:61:11)","exception":true,"date":"Thu Dec
272018 14:40:23 GMT-0200 (Brasilia Summer
Time)","process":{"pid":32071,"uid":501,"gid":20,"cwd":"/Users/biharck/Deve
loper/Hands-On-RESTful-Web-Services-with-TypeScript-3/Chapter10/order-
api","execPath":"/usr/local/bin/node","version":"v10.13.0","argv":["/usr/lo
cal/bin/node","/Users/biharck/Developer/Hands-On-RESTful-Web-Services-with-
TypeScript-3/Chapter10/order-api/node_modules/mocha/bin/_mocha","--
require","ts-node/register","--require","source-map-support/register","--
full-trace","--bail","--timeout","35000","test/**/*.spec.ts","--
reporter","spec","--compilers","ts:ts-
node/register","test/routes/01_api.spec.ts","test/routes/02_user.spec.ts","
test/routes/03_order.spec.ts","--
exit"],"memoryUsage":{"rss":176713728,"heapTotal":143851520,"heapUsed":1165
67616,"external":19504330}},"os":{"loadavg":[3.64794921875,3.7998046875,3.7
470703125],"uptime":1266247},"trace":[{"column":19,"file":"/Users/biharck/D
eveloper/Hands-On-RESTful-Web-Services-with-TypeScript-3/Chapter10/order-
api/src/controllers/order.ts","function":"order_1.OrderModel.findById","lin
e":19,"method":"findById","native":false},{"column":16,"file":"/Users/bihar
ck/Developer/Hands-On-RESTful-Web-Services-with-
TypeScript-3/Chapter10/order-
api/node_modules/mongoose/lib/model.js","function":null,"line":4645,"method
":null,"native":false},{"column":5,"file":"/Users/biharck/Developer/Hands-
On-RESTful-Web-Services-with-TypeScript-3/Chapter10/order-
api/node_modules/mongoose/lib/query.js","function":"model.Query.Query._find
One","line":1961,"method":"_findOne","native":false},{"column":33,"file":"/
Users/biharck/Developer/Hands-On-RESTful-Web-Services-with-
TypeScript-3/Chapter10/order-
api/node_modules/kareem/index.js","function":"process.nextTick","line":369,
"method":"nextTick","native":false},{"column":11,"file":"internal/process/n
ext_tick.js","function":"process._tickCallback","line":61,"method":"_tickCa
llback","native":false}],"req":{"url":"/store/orders/000","headers":{"host"
:"127.0.0.1:58682","accept-encoding":"gzip, deflate","user-agent":"node-
superagent/3.8.3","authorization":"Bearer
eyJhbGciOiJIUzI1NiIsInR5cCI6IkpXVCJ9.eyJ1c2VyIjp7Il9pZCI6IjVjMjUwMDc3MTA2Zj
Y2N2Q0NzNhOTZlNSIsImVtYWlsIjoiSm9obkBtZW1haWwuY29tIn0sImlhdCI6MTU0NTkyODgyM
30.I89VP_fVcQNwD9gxekhyYy8HUvKZia7YjoOY7ewB6wU","connection":"close"},"meth
od":"GET","httpVersion":"1.1","originalUrl":"/store/orders/000","query":{}}
}}
[2018-12-27T16:40:23.907Z] [error] => Order 000 not found.
    should respond with HTTP 404 status because there is no order (55ms)
[2018-12-27T16:40:23.911Z] [info] => [POST] [/users] { _id:
5c250077106f667d473a96e6,
  username: 'OrderUser',
  firstName: 'Order',
  lastName: 'User',
  email: 'order@myemail.com',
  password: 'password',

```
  phone: '5555555',
  userStatus: 1 }
    should create a new user for Order tests and retrieve it back (94ms)
[2018-12-27T16:40:24.010Z] [info] => [POST] [/store/orders/]
5c250077106f667d473a96e6
[2018-12-27T16:40:24.016Z] [info] => [POST] [/store/orders/] { _id:
5c250078106f667d473a96e7,
  userId: 5c250077106f667d473a96e6,
  quantity: 1,
  shipDate: 2018-12-27T16:40:22.933Z,
  status: 'PLACED',
  complete: false }
    should create a new order and retrieve it back (55ms)
[2018-12-27T16:40:24.060Z] [info] => [GET] [/store/orders/]
5c250078106f667d473a96e7
    should return the order created on the step before
[2018-12-27T16:40:24.077Z] [info] => [GET] [/store/orders/]
    should return all orders so far
[2018-12-27T16:40:24.090Z] [info] => [GET] [/store/orders/]
    should not return orders because offset is higher than the size of the
orders array
[2018-12-27T16:40:24.095Z] [info] => [GET] [/store/inventory/] PLACED
    should return the inventory for all users
[2018-12-27T16:40:24.106Z] [warn] => [DELETE] [/store/orders/]
5c250078106f667d473a96e7
    should remove an existing order
[2018-12-27T16:40:24.115Z] [warn] => [DELETE] [/store/orders/]
5c250078106f667d473a96e7
[2018-12-27T16:40:24.120Z] [warn] => [DELETE] [/store/orders/:{orderId}]
Order id 5c250078106f667d473a96e7 not found
    should return 404 when it is trying to remove an order because the
order does not exist
```

The output of the `aggregated.log` file is shown in the following screenshot:

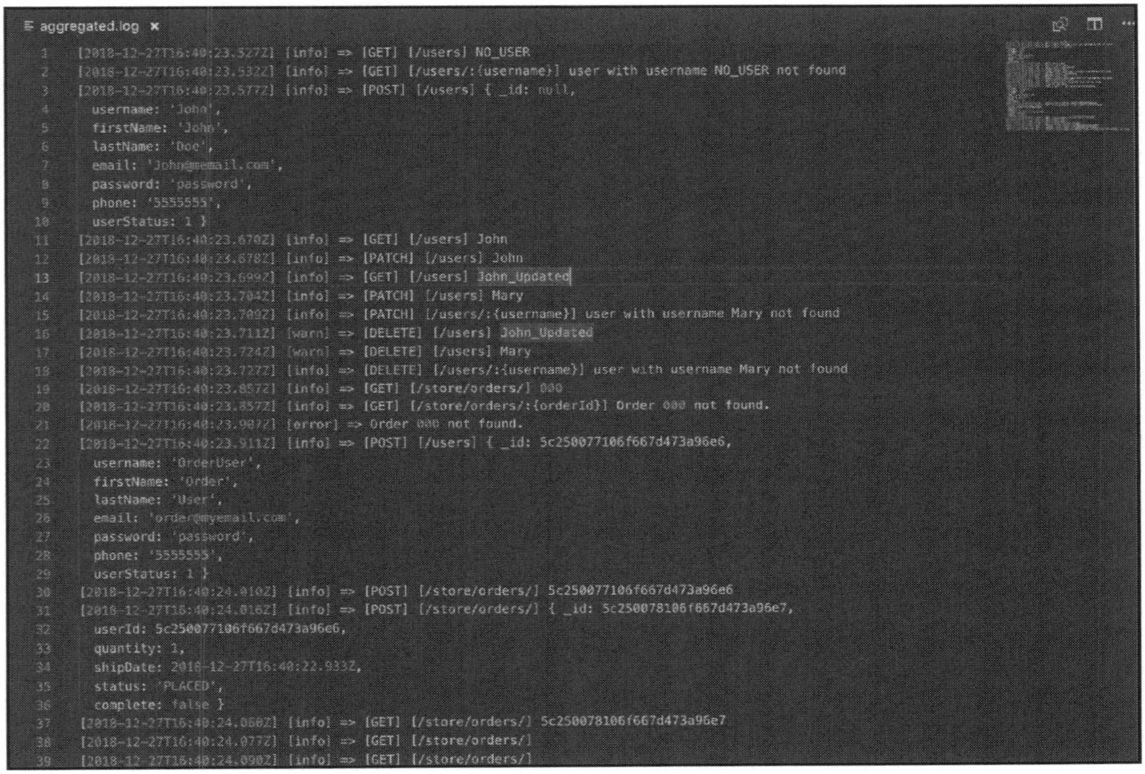

Aggregated log file output

So far, our `order-api` application is well-covered with logs, error handlers, and meaningful messages. As a challenge, take what you have learned so far and improve the logging and error handler strategies.

Summary

This chapter covered the basics of error handling and the creation of custom error handlers for different meanings, such as logging, client errors, and a default catch-all error handler. We also looked at how to create meaningful messages, not only for error messages but also for logging purposes.

A logging strategy was also presented using the `winston` library to output messages, whether on the console or in files.

Chapter 11, *Creating a CI/CD Pipeline for Your API*, introduces you to CI and CD principles, and also how to handle autoscaling with Google Cloud Platform.

Questions

1. What is the difference between an error handler and logging?
2. How do you add different kinds of error handlers to your Express.js application?
3. Is there any difference between adding an error handler definition at the beginning of an app and adding it at the end?
4. What kind of information should be logged?
5. Is there a way to add a generic logging strategy to an application?
6. What should you do if you want to grab an error message but still pass it through an error handler?
7. What does `next()` do on error handler functions?

Further reading

In order to improve your knowledge of formatting strategies, the following book is recommended, as it will be helpful in the coming chapters:

- *RESTful Web API Design with Node.js* (https://www.packtpub.com/web-development/restful-web-api-design-nodejs-10-third-edition)

Creating a CI/CD Pipeline for Your API

11

Nowadays, DevOps is a must for almost every application lifecycle. Containerization of the environment, using Continuous Integration services, running tests before deployment, and getting build notifications are the focus of this chapter. We are going to learn how to create a pipeline by Dockerizing our environment using Travis CI, Google Cloud, and GitHub.

The following topics will be covered in this chapter:

- Continuous Integration strategies
- Deployment methods
- Automated testing
- Continuous Delivery

Technical requirements

All of the information required to run the code from this chapter is provided by the chapter itself. The only necessity is to get the previous installation process from the previous chapter done, that is, Node.js, VS Code, and TypeScript.

The code base is available at `https://github.com/PacktPublishing/Hands-On-RESTful-Web-Services-with-TypeScript-3/tree/master/Chapter11`.

Continuous Integration

Continuous Integration (CI) is a development practice that involves developers integrating their code into a shared repository several times per day. Every time that code is sent to a repository, an automated process runs several processes, such as an automated build, testing, and other tasks, helping developers to identify possible problems before going to production.

The practice of integrating regularly allows developers to detect errors sooner than with normal processes, which, of course, means that they can be solved sooner too.

To exemplify this concept, this chapter will walk you through a process that involves Travis CI being triggered when changes are pushed to GitHub. It will then run tests automatically, and, if the tests pass, it will deploy a new version of order-api to Google Cloud Platform:

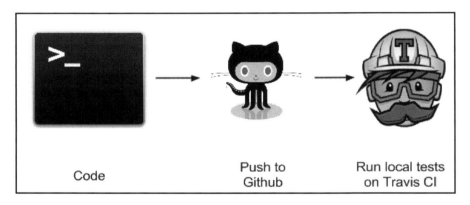

The Continuous Integration process with Travis CI - adapted from https://cloud.google.com/solutions/continuous-delivery-with-travis-ci

Adding the code to GitHub

Before we get started, we have to add our code to a GitHub repository if it is not already there. Go to your GitHub account (`https://github.com/`) and create a new repository called `order-api`:

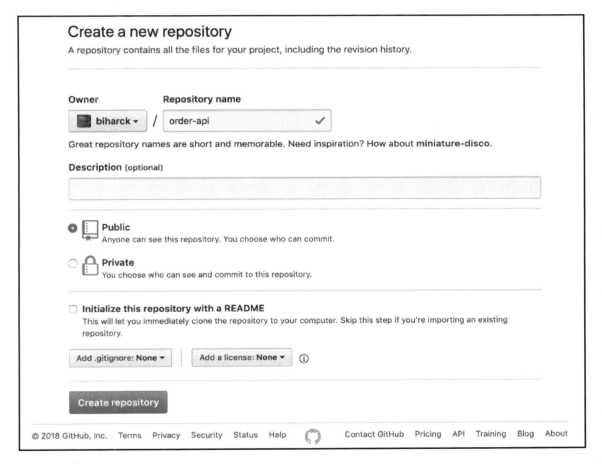

Creating a new repository called order-api on GitHub

After that, add a file called `.gitignore` to the root folder so that unnecessary and undesirable files don't go to the repository:

```
node_modules/
package-lock.json
.DS_Store
package-lock.json
dist
.stryker-tmp
reports
launch.json
*.pem
*.log
keys.json
gce.json
*.env
*.prod
```

Then, on Terminal into the root folder of your project, run the following command to create a new `readme` file:

```
$ echo "# order-api" >> README.md
```

After that, initialize GitHub and add the files to be pushed to the remote repository:

```
$ git init
$ git add .
$ git commit -m "adding order api files to the github"
```

You should see output more or less like this:

```
[master (root-commit) 0c7042b] adding order api files to the github
 29 files changed, 1005 insertions(+)
 create mode 100644 .gitignore
 create mode 100644 .prettierrc
 create mode 100644 README.md
 create mode 100644 package.json
 create mode 100644 src/app.ts
 create mode 100644 src/controllers/api.ts
 create mode 100644 src/controllers/order.ts
 create mode 100644 src/controllers/user.ts
 create mode 100644 src/models/applicationType.ts
 create mode 100644 src/models/order.ts
 create mode 100644 src/models/orderStatus.ts
 create mode 100644 src/models/user.ts
 create mode 100644 src/routes/api.ts
 create mode 100644 src/routes/order.ts
 create mode 100644 src/routes/user.ts
```

```
create mode 100644 src/schemas/order.ts
create mode 100644 src/schemas/user.ts
create mode 100644 src/server.ts
create mode 100644 src/utility/errorHandler.ts
create mode 100644 src/utility/logger.ts
create mode 100644 src/utility/orderApiUtility.ts
create mode 100644 src/utility/passportConfiguration.ts
create mode 100644 stryker.conf.js
create mode 100644 test/mocha.opts
create mode 100644 test/routes/01_api.spec.ts
create mode 100644 test/routes/02_user.spec.ts
create mode 100644 test/routes/03_order.spec.ts
create mode 100644 tsconfig.json
create mode 100644 tslint.json
```

Finally, create the relationship between your local code and the remote repository:

```
$ git remote add origin git@github.com:<YOUR_GITHUB_USER>/order-api.git
```

The final command sends the code to GitHub:

```
$ git push -u origin master
```

The output might be as follows:

```
Counting objects: 39, done.
Delta compression using up to 4 threads.
Compressing objects: 100% (38/38), done.
Writing objects: 100% (39/39), 10.34 KiB | 2.07 MiB/s, done.
Total 39 (delta 4), reused 0 (delta 0)
remote: Resolving deltas: 100% (4/4), done.
To github.com:<YOUR_GIT_ACCOUNT>/order-api.git
 * [new branch] master -> master
Branch master set up to track remote branch master from origin.
```

The code will then be available at the GitHub repository:

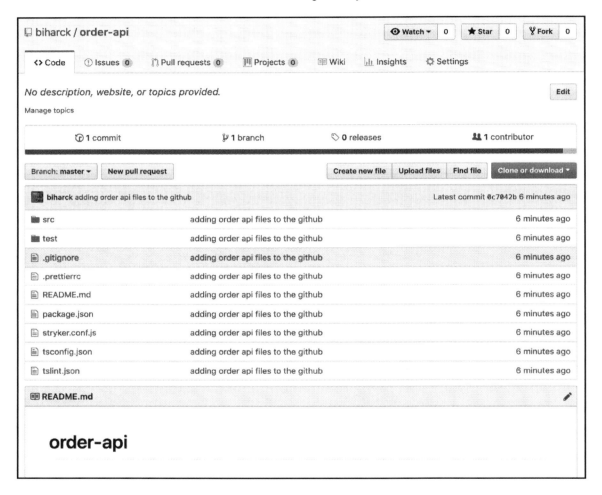

The order-api code base pushed to GitHub

Continuous Integration with Travis CI

Travis CI is a Cloud Integration web service integrated with GitHub. It is free for public repositories (`https://travis-ci.org/`) and paid for private repositories `https://travis-ci.org/`. It was developed in Ruby and its components are distributed under the MIT license:

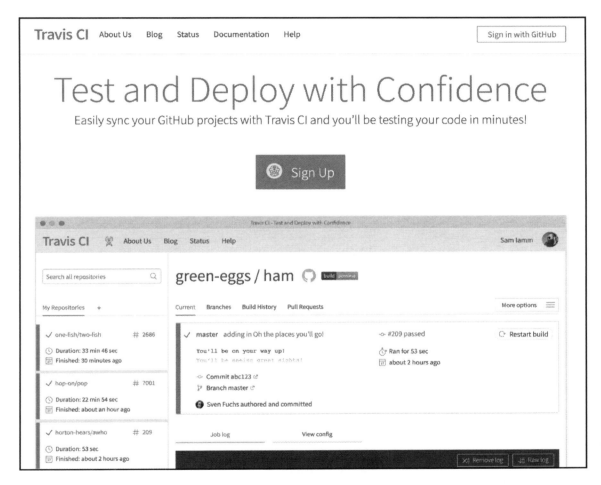

Travis CI home page

If you do not have a Travis CI account yet, go to `https://travis-ci.org/` and create one for your public project. It is easy—you just have to sign in to your GitHub account and it is done.

Right after you create the account, log in and look at the dashboard. You will notice that there is no project associated with Travis CI yet. In order to add `order-api` to the list of Travis CI repositories, go to `https://travis-ci.org/account/repositories`, type `order-api` under **Legacy Services Integration**, and flag the repository as selected on the left-hand side:

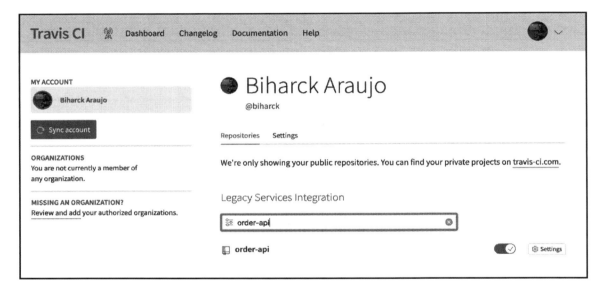

Travis CI repository list

From now on, Travis CI will be able to listen to changes on this repository and, once we have our Travis CI files properly configured, we will be able to see Travis CI running `order-api` jobs there.

Go back to your project and create a file called `.travis.yml` in the root folder, with the following content:

```
language: node_js
node_js:
  - '10'
sudo: required
services:
  - docker
script:
  - npm install
```

Commit this file and push the changes to GitHub:

```
$ git add .travis.yml
$ git commit -m "adding travis file"
$ git push
```

Go to the Travis CI dashboard and click on the `order-api` project to see the running pipeline:

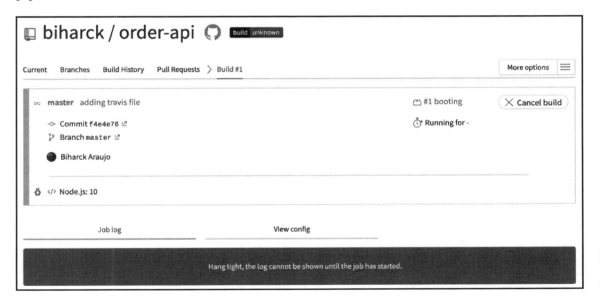

The pipeline is triggered based on the previous push to GitHub

When the pipeline finishes, the color changes to green if it passes and red if it fails:

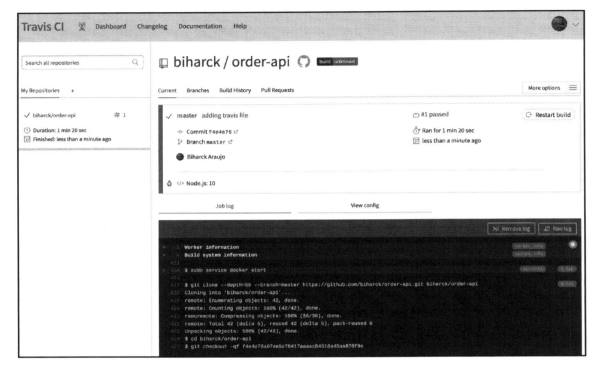

Pipeline passed on Travis CI

Basically, due to our `.travis.yml` file, we asked Travis to run the `npm install` command and it ran successfully. Now, every time you push changes to GitHub, Travis will trigger a new pipeline.

Adding tests to the pipeline

Now that we have the pipeline in place, let's add a more meaningful process, such as integration testing. Since almost everything is in place, we just have to include two more pieces of information in our `.travis.yml` file—the MongoDB service and the `test` command:

```
language: node_js
node_js:
  - '10'
sudo: required
```

```
services:
  - docker
  - mongodb
script:
  - npm run test
```

Since we are going to handle more than one environment, that is, dev, test, and prod with different information, we should start working with a tool to help us handle environment variables. To do that, we will install the dotenv library:

```
$ npm install --save dotenv
```

Also, there are some adjustments to be made to the package.json file, because we are going to use the order-api application in production mode and there are some libraries that are only dev dependencies that should also be prod dependencies. The new package.json file is as follows:

```
{
  "name": "order-api",
  "version": "1.0.0",
  "description": "This is the example from the Book Hands on RESTful APIs
with TypeScript 3",
  "main": "./dist/server.js",
  "scripts": {
    "build": "npm run clean && tsc",
    "clean": "rimraf dist && rimraf reports",
    "lint": "tslint ./src/**/*.ts ./test/**/*.spec.ts",
    "lint:fix": "tslint --fix ./src/**/*.ts ./test/**/*.spec.ts -t
verbose",
    "pretest": "cross-env NODE_ENV=test npm run build && npm run lint",
    "test": "cross-env NODE_ENV=test mocha --reporter spec --compilers
ts:ts-node/register test/**/*.spec.ts --exit",
    "test:mutation": "stryker run",
    "stryker:init": "stryker init",
    "dev": "cross-env PORT=3000 NODE_ENV=dev ts-node ./src/server.ts",
    "prod": "PORT=3000 NODE_ENV=prod npm run build && npm run start",
    "start": "NODE_ENV=prod node dist/server.js",
    "tsc": "tsc"
  },
  "engines": {
    "node": ">=8.0.0"
  },
  "keywords": [
    "order POC",
    "Hands on RESTful APIs with TypeScript 3",
    "TypeScript 3",
    "Packt Books"
```

```
  ],
  "author": "Biharck Muniz Araújo",
  "license": "MIT",
  "devDependencies": {
    "@types/bcrypt": "^3.0.0",
    "@types/body-parser": "^1.17.0",
    "@types/chai": "^4.1.7",
    "@types/chai-http": "^3.0.5",
    "@types/express": "^4.16.0",
    "@types/express-winston": "^3.0.0",
    "@types/jsonwebtoken": "^8.3.0",
    "@types/mocha": "^5.2.5",
    "@types/mongodb": "^3.1.17",
    "@types/mongoose": "^5.3.5",
    "@types/passport": "^0.4.7",
    "@types/passport-jwt": "^3.0.1",
    "@types/passport-local": "^1.0.33",
    "@types/mongoose-unique-validator": "^1.0.1"
  },
  "dependencies": {
    "@types/node": "^10.12.12",
    "bcrypt": "^3.0.2",
    "body-parser": "^1.18.3",
    "chai": "^4.2.0",
    "chai-http": "^4.2.0",
    "cross-env": "^5.2.0",
    "dotenv": "^6.2.0",
    "express": "^4.16.4",
    "express-winston": "^3.0.1",
    "halson": "^3.0.0",
    "js2xmlparser": "^3.0.0",
    "lodash": "^4.17.11",
    "mocha": "^5.2.0",
    "mongoose": "^5.4.0",
    "mongoose-unique-validator": "^2.0.2",
    "passport": "^0.4.0",
    "passport-jwt": "^4.0.0",
    "passport-local": "^1.0.0",
    "rimraf": "^2.6.2",
    "stryker": "^0.33.1",
    "stryker-api": "^0.22.0",
    "stryker-html-reporter": "^0.16.9",
    "stryker-mocha-framework": "^0.13.2",
    "stryker-mocha-runner": "^0.15.2",
    "stryker-typescript": "^0.16.1",
    "ts-node": "^7.0.1",
    "tslint": "^5.11.0",
    "tslint-config-prettier": "^1.17.0",
```

```
      "typescript": "^3.2.1",
      "winston": "^3.1.0"
  }
}
```

Also, go to the `app.js` file, remove the `../src/` folder reference from the routes, and change the way we now get MongoDB credentials using environment variables such as the following:

```
import * as bodyParser from 'body-parser'
import * as dotenv from 'dotenv'
import * as express from 'express'
import * as expressWinston from 'express-winston'
import * as mongoose from 'mongoose'
import * as winston from 'winston'
import { APIRoute } from './routes/api'
import { OrderRoute } from './routes/order'
import { UserRoute } from './routes/user'
import * as errorHandler from './utility/errorHandler'

class App {
  public app: express.Application
  public userRoutes: UserRoute = new UserRoute()
  public apiRoutes: APIRoute = new APIRoute()
  public orderRoutes: OrderRoute = new OrderRoute()
  public mongoUrl: string
  public mongoUser: string
  public mongoPass: string

  constructor() {
    const path = `${__dirname}/../.env.${process.env.NODE_ENV}`
    dotenv.config({ path: path })
    this.mongoUrl = `mongodb://${process.env.MONGODB_URL_PORT}/${
      process.env.MONGODB_DATABASE
    }`
    this.mongoUser = `${process.env.MONGODB_USER}`
    this.mongoPass = `${process.env.MONGODB_PASS}`

    this.app = express()
    this.app.use(bodyParser.json())
    this.userRoutes.routes(this.app)
    this.apiRoutes.routes(this.app)
    this.orderRoutes.routes(this.app)
    this.mongoSetup()
    this.app.use(
      expressWinston.errorLogger({
        transports: [new winston.transports.Console()],
      })
```

```
    )
    this.app.use(errorHandler.logging)
    this.app.use(errorHandler.clientErrorHandler)
    this.app.use(errorHandler.errorHandler)
  }

  private mongoSetup(): void {
    let options

    if (process.env.NODE_ENV !== 'prod') {
      options = {
        useNewUrlParser: true,
      }
    } else {
      options = {
        user: this.mongoUser,
        pass: this.mongoPass,
        useNewUrlParser: true,
      }
    }
    mongoose.connect(
      this.mongoUrl,
      options
    )
  }
}

export default new App().app
```

As you can see, we now have to have environment variable files to help us with configurations. Up to now, we just needed the dev and test ones. Create two files—one called .env.dev in the root folder, with the following content:

```
MONGODB_URL_PORT=localhost:27017
MONGODB_DATABASE=order-api
```

Then, create another one called .env.test with the following content:

```
MONGODB_URL_PORT=localhost:27017
MONGODB_DATABASE=order-api-test
```

Now, just add the files, and commit and push the changes. Go to Travis CI and follow the pipeline. You will see that the tests will run on Travis CI. This means that every change that is pushed to the master branch will trigger the Travis CI pipeline, which will run the integration tests:

```
549      ✓ should respond with HTTP 404 status because there is no order
550  [2018-12-27T19:52:37.794Z] [info] => [POST] [/users] { _id: 5c252d85bb50be11ffb1ccb2,
551    username: 'OrderUser',
552    firstName: 'Order',
553    lastName: 'User',
554    email: 'order@myemail.com',
555    password: 'password',
556    phone: '5555555',
557    userStatus: 1 }
558      ✓ should create a new user for Order tests and retrieve it back (83ms)
559  [2018-12-27T19:52:37.877Z] [info] => [POST] [/store/orders/] 5c252d85bb50be11ffb1ccb2
560  [2018-12-27T19:52:37.880Z] [info] => [POST] [/store/orders/] { _id: 5c252d85bb50be11ffb1ccb3,
561    userId: 5c252d85bb50be11ffb1ccb2,
562    quantity: 1,
563    shipDate: 2018-12-27T19:52:37.158Z,
564    status: 'PLACED',
565    complete: false }
566      ✓ should create a new order and retrieve it back
567  [2018-12-27T19:52:37.897Z] [info] => [GET] [/store/orders/] 5c252d85bb50be11ffb1ccb3
568      ✓ should return the order created on the step before
569  [2018-12-27T19:52:37.902Z] [info] => [GET] [/store/orders/]
570      ✓ should return all orders so far
571  [2018-12-27T19:52:37.907Z] [info] => [GET] [/store/orders/]
572      ✓ should not return orders because offset is higher than the size of the orders array
573  [2018-12-27T19:52:37.910Z] [info] => [GET] [/store/inventory/] PLACED
574      ✓ should return the inventory for all users
575  [2018-12-27T19:52:37.915Z] [warn] => [DELETE] [/store/orders/] 5c252d85bb50be11ffb1ccb3
576      ✓ should remove an existing order
577  [2018-12-27T19:52:37.919Z] [warn] => [DELETE] [/store/orders/] 5c252d85bb50be11ffb1ccb3
578  [2018-12-27T19:52:37.920Z] [warn] => [DELETE] [/store/orders/:{orderId}] Order id 5c252d85bb50be11ffb1ccb3 not found
579      ✓ should return 404 when it is trying to remove an order because the order does not exist
580
581
582    21 passing (762ms)
583
584  The command "npm run test" exited with 0.
585
586
587
588  Done. Your build exited with 0.
```

Travis CI output of the tests

Continuous Deployment

Considering that we can now see the tests running every push to the master branch, we can assume that all green builds are candidates for being deployed, which is basically the high-level definition of Continuous Deployment.

In this section, you will learn how to deploy your application to **Google Cloud Platform** (**GCP**) manually and automatically with Travis CI.

Manual deployment to GCP

Before we move on, you must have a Google Cloud account already set up. Go to `http://cloud.google.com` and click **Try it Free** to create an account and start a free trial.

Once you've created your account, go to the project list and **Create a project** and name it `order-api`. After you've finished, you'll probably be redirected to the Google Cloud Platform **Dashboard**:

Google Cloud Platform Dashboard

In order to continue, you have to set a billing account. Go to `http://console.cloud.google.com/billing/projects` to do that. Once you finish setting up, the page should look like this:

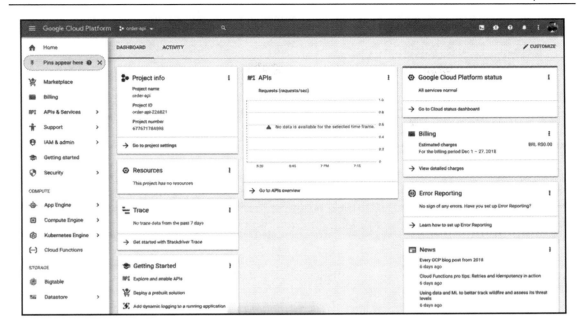

Google Cloud Platform after a billing account is set up

The next step is to install the Google Cloud CLI, also know as the `gcloud` CLI, available at `http://cloud.google.com/sdk`. Go there and install the `gcloud` CLI:

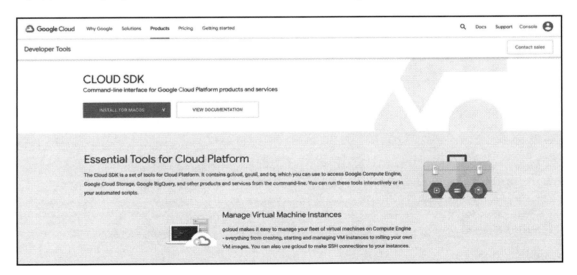

The gcloud SDK home page

The `gcloud` CLI documentation is available at `https://cloud.google.com/sdk/docs/`.

When you have `gcloud` installed, run the following command to connect the CLI to the project you created:

```
$ gcloud init
```

When you run this command, Google will take you to its multi-factor authorization flow, which will create an API key that will be used by the CLI to make authenticated API requests to your Google Cloud Platform account. Right after you finish the authentication step, run the following command to validate that you've set the correct project:

```
$ gcloud info | grep project:
```

The output should be similar to the following:

```
project: [order-api-226821]
```

No worries about the number after the second dash, because it probably won't be this one. This is the project ID generated based on your project name.

Since we now have the account created and properly set up, we can start the manual deployment. To do so, create a new file called `app.yaml` in the root `order-api` folder, with the following content:

```
runtime: custom
env: flex

manual_scaling:
  instances: 1
resources:
  cpu: 1
  memory_gb: 0.5
  disk_size_gb: 10
```

This means that Google will use the custom indicator to configure your environment entirely with Docker. Also, we are setting up manual scaling to reduce costs since this is just a test for now. There is just one instance with 1 CPU and 500 MB of memory, also with 10 GB of disk size.

Of course, before we move on, we must have a Docker for our application. To do that, create a new file called Dockerfile under the same level as app.yaml, with the following content:

```
FROM node:10

ENV NODE_ENV production
# Create app directory
RUN mkdir -p /usr/src/app
WORKDIR /usr/src/app

# Install app dependencies
COPY . .

RUN npm install -g typescript
RUN npm install
RUN npm install --dev


EXPOSE 3000

CMD ["npm", "run", "prod"]
```

Also, create a file called .dockerignore on the same level as the content:

```
node_modules
```

This is just to ignore the node_modules folder when building a new image.

We also have to remove our HTTPS configuration from the server.ts file, since GCP is going to handle it for us automatically:

```
import app from './app'

const PORT = process.env.PORT || 3000

app.listen(PORT)
```

Then, run the build command to build a new Docker image for the order-api application:

```
$ docker build -t order-api
```

When the `build` command finishes, run the following command to test it out locally with Docker:

```
$ docker run -d --name myMongoDB mongo
```

The following command links MongoDB to the localhost:

```
$ docker run --link=myMongoDB:mongodb -p 3000:3000 -it order-api
```

These commands will spin up two Docker containers, one for `order-api` and another one for MongoDB. This command will also link them both so that the `order-api` application will be able to see the MongoDB container as `localhost`.

To test it, run the tests under Docker:

```
$ npm run test
```

If the tests pass, everything is fine, and we can move on and push the changes to GitHub:

```
$ git add .
$ git commit -m "adding google cloud configuration"
$ git push
```

Check whether the pipeline passes or not. If it is green, we can move on and deploy the first version to GCP with this command:

```
$ gcloud app deploy
```

 The first time, this deploy process takes several minutes.

The command will upload the `order-api` application to the container registry. After that, it will tell Google App Engine to create and start the first service of the project as `default`:

```
target project: [order-api-226821]
target service: [default]
target version: [20181227t195707]
target url: [https://order-api-226821.appspot.com]
```

Notice that the target URL will be `https://order-api-226821.appspot.com`. The information message when deploying, is as follows:

```
Creating App Engine application in project [order-api-226821] and region
[us-east1]....
```

When it finishes, the output will be as follows:

```
Updating service [default] (this may take several minutes)...done.
Setting traffic split for service [default]...done.
Deployed service [default] to [https://order-api-226821.appspot.com]

You can stream logs from the command line by running:
  $ gcloud app logs tail -s default

To view your application in the web browser run:
  $ gcloud app browse
```

Go to the target URL to see whether the app is running:

```
https://order-api-226821.appspot.com/api
```

The output should be the following:

```
{
    "title": "Order API"
}
```

If you try to reach any other endpoint, you will see that an error will appear. That's because we never set a MongoDB instance on GCP. To do that, there are a few possibilities, as described in Google's documentation:

- A Google Compute Engine virtual machine has to be created. Also ensure that MongoDB is pre-installed on that machine.
- A MongoDB instance has to be created. This instance should have MongoDB Atlas on GCP.
- Create a free MongoDB deployment using mLab on Google Cloud Platform.

We will go with the third option, mLab, since there is a free quota there we could use to test our application.

Configuring mLab

As a MongoDB cloud database service, mLab is a third-party service that runs on cloud providers. It has also collaborated with PaaS providers:

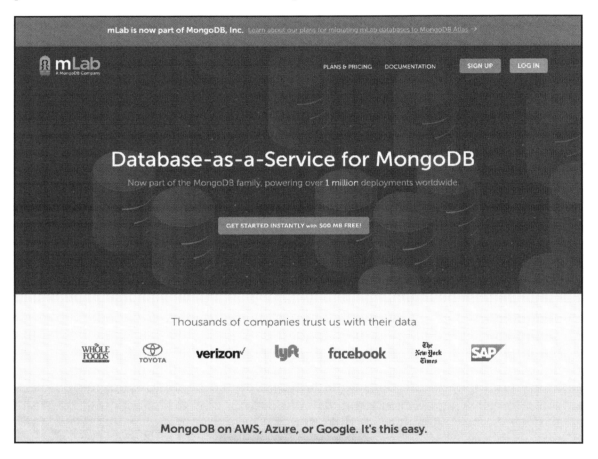

mLab home page

mLab provides a free sandbox version up to 0.5 GB, which is enough for our `order-api` application. Go to the mLab website (`https://mlab.com/`) and create a new account.

After you create the account, create a new database for Google Cloud Platform, called `order-api`:

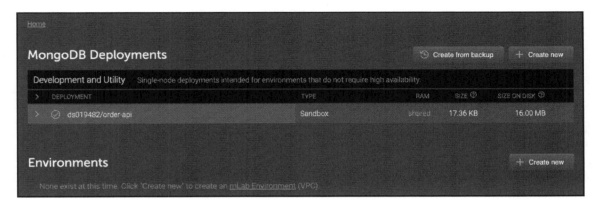

Once you've created the database, create a new user for this database, called `order-api-user`, with the password as `handsOnTypeScript3`.

Right after the tests, change the password to a stronger, more secure one.

And that's it! We now have a MongoDB account with a user to access it. To see how to connect to this database, check the top of the page for a disclaimer like this:

```
To connect using the mongo shell:
mongo ds019482.mlab.com:19482/order-api -u <dbuser> -p <dbpassword>
To connect using a driver via the standard MongoDB URI (what's this?):

mongodb://<dbuser>:<dbpassword>@ds019482.mlab.com:19482/order-api
```

The next section will walk you through how to deploy the application on GCP, and also how to set production secrets for MongoDB.

Automatic deployment to GCP

After all the previous configuration, we are now able to apply the automatic deployment or Continuous Delivery practices to our application. The idea now is that, for every green build, which means that the tests passed, we will deploy this new version to production through Travis CI, as in the following diagram:

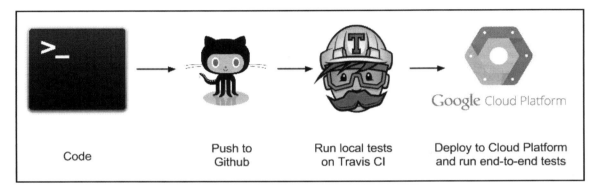

| Code | Push to Github | Run local tests on Travis CI | Deploy to Cloud Platform and run end-to-end tests |

The Continuous Delivery process with GitHub, Travis CI, and GCP. Source: https://cloud.google.com/solutions/continuous-delivery-with-travis-ci

Luckily, almost all the configurations are already in place and we just have to change the `.travis.yml` file a little bit to include the deployment step and include permissions on GCP for that.

Let's get started adding the rights to GCP. To do so, go to the **Service accounts** page by typing `service account` in the search bar:

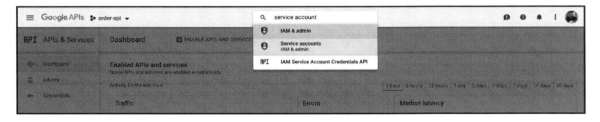

Searching for the Service account page

You will then see the **Service accounts** page, as shown:

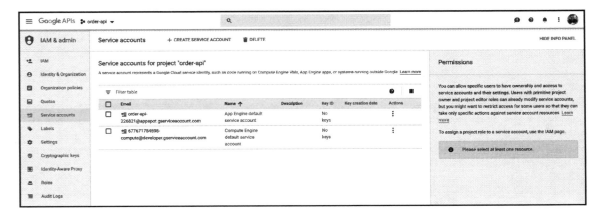

Service account page

Click on the **+ CREATE SERVICE ACCOUNT** link and fill the form using `Travis Deploy` as the **Service account name**:

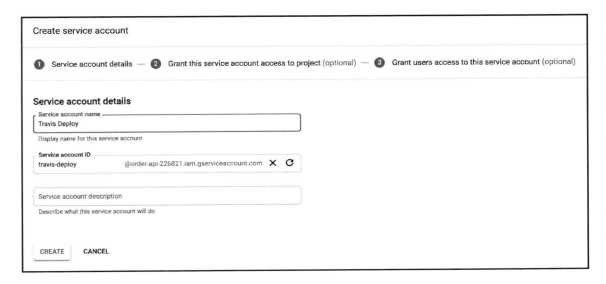

Service account creation page

Then, select **Project** | **Editor** as the role:

Service account role

Then, select **+ CREATE KEY** and select the **JSON** radio button:

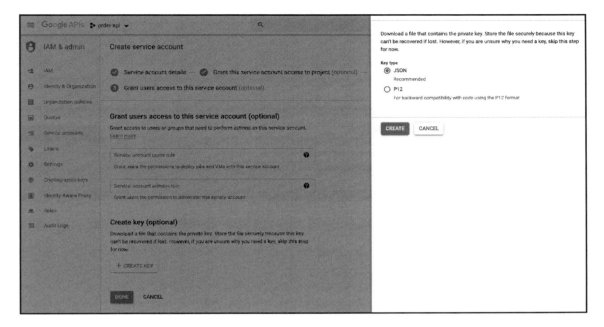

Service account key

When you click on **CREATE**, a key will be downloaded to your local machine. Be careful with that. Do not push this key to any repository or share it. Do not forget to click on **DONE**. Lastly, in the search bar, you have to type App Engine Admin API and then click on **ENABLE**:

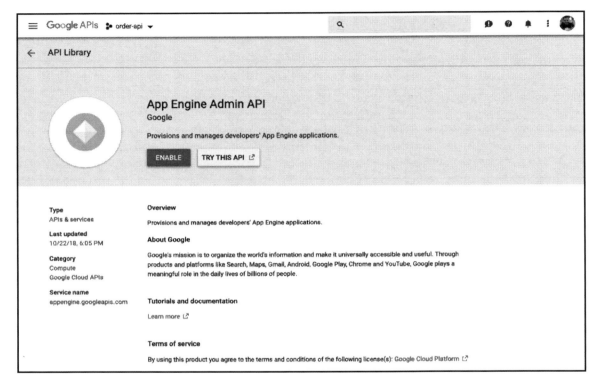

Enabling the App Engine Admin API

Now, go to your .travis.yml file and add the deployment scope. Change the project value to your GCP project ID:

```
language: node_js
node_js:
  - '10'
sudo: required
services:
  - docker
  - mongodb
script:
  - npm run test
deploy:
```

```
provider: gae
project: order-api-226821
keyfile: gce.json
skip_cleanup: true
verbosity: debug
on: master
```

 Remember to change the project value on the deployment scope with your project ID.

The new information added is as follows:

- provider: gae: Use the Google App Engine deploy target
- project: order-api-226821: Your GCP Project ID
- keyfile: gce.json: The deploy key to use
- verbosity: debug: The log level
- on: master: Only deploy to GCP commits on master

Now, you have to encrypt the key you download from GCP and add the encrypted value on Travis CI. To do that, first, move the key to your root folder and rename it gce.json.

Then, add this file to the .gitignore file to make sure you will not push this key to GitHub:

```
$ echo "gce.json" >> .gitignore
```

Encrypt gce.json by running travis encrypt-file:

```
travis encrypt-file gce.json --add
encrypting gce.json for biharck/order-api
storing result as gce.json.enc
storing secure env variables for decryption

Make sure to add gce.json.enc to the git repository.
Make sure not to add gce.json to the git repository.
Commit all changes to your .travis.yml.
```

With this command, a new file called `gce.json.enc` is created. This command also modifies the content of `.travis.yml` by adding data as shown:

```
before_install:
- openssl aes-256-cbc -K $encrypted_5c02ebcbdbd5_key -iv
$encrypted_5c02ebcbdbd5_iv
   -in gce.json.enc -out gce.json -d
```

So, your full `.travis.yml` file right now is as follows:

```
language: node_js
node_js:
  - '10'
sudo: required
services:
  - docker
  - mongodb
script:
  - npm run test
deploy:
  provider: gae
  project: order-api-226821
  keyfile: gce.json
  skip_cleanup: true
  verbosity: debug
  on: master
before_install:
  - openssl aes-256-cbc -K $encrypted_5c02ebcbdbd5_key -iv
$encrypted_5c02ebcbdbd5_iv
    -in gce.json.enc -out gce.json -d
```

So far, we are good to deploy the changes to production, but we still have to add the production credentials for MongoDB. Since we already changed our `app.ts` file to support environment variables, we could just create a new `.env.prod` file with the production credentials, but this is not secure at all since the credentials will be available at the repository.

What we are going to do is add the credentials to Google Cloud Storage and, at runtime, the application available on App Engine will download the `.env.prod` file from Google's storage bucket and apply it.

In order to use Google Cloud Storage, Go to **Google Cloud Platform** | **Storage** | **Browser** and create a new bucket there:

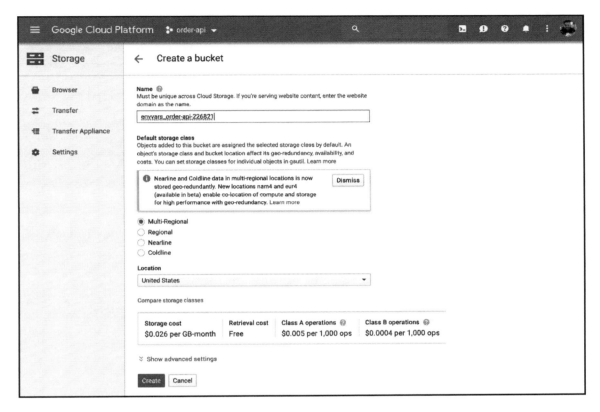

Creating a new bucket on Google Cloud Storage

When you finish creating the bucket, upload a file called `.env.prod` with the content you got from mLab to this new bucket:

```
MONGODB_URL_PORT=ds019482.mlab.com:19482
MONGODB_DATABASE=order-api
MONGODB_USER=order-api-user
MONGODB_PASS=handsOnTypeScript3
```

Note that the `.env.prod` file contains sensitive information and should not be pushed to GitHub.

.env.prod file uploaded to the new bucket

To summarize, we have a new bucket on Google Cloud Storage with the production credentials. Also, our `app.ts` is prepared to get environment variables for any environment, including production, but we still have to tell the application that every time that production mode is activated, we want to get the information from the `.env.prod` file.

To do so, include the `prestart` task in the `package.json` file, like so:

```
"dev": "cross-env PORT=3000 NODE_ENV=dev ts-node ./src/server.ts",
"prod": "PORT=3000 NODE_ENV=prod npm run build && npm run start",
"prestart": "node dist/utility/config.js",
"start": "NODE_ENV=prod node dist/server.js",
```

This means that every time that the `start` task is called, first the `prestart` will be called, and the responsibility of that task is to grab the production environment variable.

As you will notice, the `prestart` task is calling a file called `config.ts`, so let's create this file under `src/utility/config.ts` with the following content:

```
import { Storage } from '@google-cloud/storage'

const downloadFile = (
  bucketName: string,
  srcFilename: string,
  destFilename: string
): void => {
  const storage = new Storage()
```

```
const options = {
  destination: destFilename,
}

storage
  .bucket(bucketName)
  .file(srcFilename)
  .download(options)
}

downloadFile('envvars_order-api', '.env.prod', '.env.prod')
```

Basically, what this file does is download the env file to the order-api project. Note that there is a new library called @google-cloud/storage, which handles communication with Google Cloud Storage. Install this library as follows:

```
$ npm install --save @google-cloud/storage
```

Finally, add the files, commit and push them, and wait for the pipeline to complete. When that's done, the new version should be deployed to GCP automatically.

> If you want to be notified every time the pipeline runs, go to the Travis CI settings and enable email notifications.
> Remember that the deployment process takes a while.

After the deployment, complete as the following screenshot:

```
[2018-12-29T02:09:26.769Z] [info] => [GET] [/store/orders/]
    ✓ should not return orders because offset is higher than the size of the orders array
[2018-12-29T02:09:26.775Z] [info] => [GET] [/store/inventory/] PLACED
    ✓ should return the inventory for all users
[2018-12-29T02:09:26.781Z] [warn] => [DELETE] [/store/orders/] 5c26d7566d7216129d22bb3e
    ✓ should remove an existing order
[2018-12-29T02:09:26.787Z] [warn] => [DELETE] [/store/orders/] 5c26d7566d7216129d22bb3e
[2018-12-29T02:09:26.789Z] [warn] => [DELETE] [/store/orders/:{orderId}] Order id 5c26d7566d7216129d22bb3e not found
    ✓ should return 404 when it is trying to remove an order because the order does not exist


  21 passing (685ms)

The command "npm run test" exited with 0.


$ rvm $(travis_internal_ruby) --fuzzy do ruby -S gem install dpl

Installing deploy dependencies
Preparing deploy
Deploying application
Done. Your build exited with 0.
```

Application deployed successfully

Connect to mLab using Robomongo and create a new user there manually with the following content:

```
{
    "username" : "John",
    "firstName" : "John",
    "lastName" : "Doe",
    "email" : "unique_email@email.com",
    "password" :
"$2b$10$GFS8IF2KL7CNdl6mJ/zNRu93j4pFDtRwTiaG73GUOJo.verzFMxQ.",
    "phone" : "5555555",
    "userStatus" : 1
}
```

This will allow you to log in to the production application for the first time, since the database is empty:

Call GET using this URL: `https://order-api-226821.appspot.com/users/login?username=Johnpassword=password`.

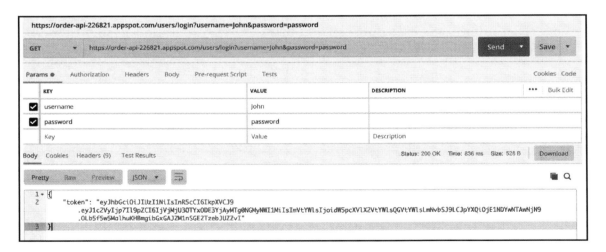

Getting the token from the production application

Since the application is now in production, we should consider adding a duration (expires in) for the JWT token. See the documentation at `https://www.npmjs.com/package/jsonwebtoken`.

Congratulations! We now have Continuous Deployment in place, so every commit with a green build is going to be available in production automatically.

Scalability

As described on the Google documentation page, the keys used to control the scaling of a service depend on the type of scaling you assign to the service. You can use either automatic or manual scaling. The default is automatic scaling.

You can configure automatic scaling by adding an `automatic_scaling` section to your `app.yaml` file, like this:

```
automatic_scaling:
  min_num_instances: 1
  max_num_instances: 15
  cool_down_period_sec: 180
  cpu_utilization:
    target_utilization: 0.6
```

This means that autoscaling is now automatic with a minimum of one instance given to the `order-api` application, and with a maximum of `15` instances. As described in Google's documentation, the `cool_down_period_sec` means the minimum number of seconds that the autoscaler should wait after which it can start gathering data from a new instance. If the autoscaler gathers data from an instance while it's still initializing, such data is not going to be reliable. Hence, the cool-down period stops the autoscaler from gathering any data during the initial period. The default cool-down period is 120 seconds, however, it must be greater than or equal to 60 seconds.

Coming to `target_utilization`, first the average CPU use is determined from all running instances. Then, it's used to determine whether the number of instances should be increased or reduced. The default value is 0.5. It's worth noticing that, once a shutdown signal is received by an instance, the instances are downscaled after 25 seconds, irrespective of requests coming in-flight.

 Be careful! This will increase costs a lot.

Summary

This chapter walked you through some Continuous Integration and Continuous Delivery techniques, where you were able to automate testing, building, and deployment processes, making sure that every new, green build goes to production without human intervention.

You also learned the basics of GCP, Travis CI, and mLab for MongoDB, making your API ready for production with to deployment methods and deployment services.

Chapter 12, *Developing RESTful APIs with Microservices*, introduces you to the principles of microservices and how we could split our order-api into microservices.

Questions

1. What does Continuous Integration mean?
2. What does Continuous Delivery mean?
3. What is Travis CI?
4. What is GCP?
5. What are the possibilities when using MongoDB on GCP?
6. Tell at least on information that shouldn't you push to production environment?
7. What happened if a build is not green on Travis CI with CD?

Further reading

In order to improve your knowledge about formatting strategy, the following books are recommended, as they will be helpful in the coming chapters:

- *Git – Version Control for Everyone* (https://www.packtpub.com/application-development/git-version-control-everyone)
- *Hands-On Continuous Integration and Delivery* (https://www.packtpub.com/virtualization-and-cloud/hands-continuous-integration-and-delivery)
- *DevOps: Continuous Delivery, Integration, and Deployment with DevOps* (https://www.packtpub.com/virtualization-and-cloud/devops-continuous-delivery-integration-and-deployment-devops)

- *Google Cloud Platform Administration* (`https://www.packtpub.com/virtualization-and-cloud/google-cloud-platform-administration`)
- *Google Cloud Platform for Developers* (`https://www.packtpub.com/virtualization-and-cloud/google-cloud-platform-developers`)

4
Section 4: Extending the Capabilities of RESTful Web Services

In this section, you will learn how to develop RESTful microservices, use GraphQL, and extend the concept to the cloud and the IoT.

The following chapters are included in this section:

- Chapter 12, *Developing RESTful APIs with Microservices*
- Chapter 13, *Flexible APIs with GraphQL*

Developing RESTful APIs with Microservices

12

When you have your own RESTful API, you will start to think about how to run each service independently and control them inside their own medium. Microservices is a hot topic nowadays and we will start from the definition of what a microservice is and is not. Then, we will continue on to the isolation of APIs within the environment so that they can run autonomously. We will also explore the possibilities for splitting an existing API into a smaller, more scalable microservice.

The following topics will be covered in this chapter:

- What are microservices?
- Isolation and resilience
- Scalability
- API Portal

Technical requirements

All of the information required to run the code from this chapter is provided by the chapter itself. The only necessity is to get the installation process from the previous chapter done, that is, Node.js, VS Code, and TypeScript.

The code base is available at `https://github.com/PacktPublishing/Hands-On-RESTful-Web-Services-with-TypeScript-3/tree/master/Chapter12`.

What are microservices?

Being an architectural style, microservices, or microservice architecture, carry out structuring a solution as a collection of services that are tightly coupled. This architecture also facilitates the delivery and **Continuous Deployment (CD)** of large, complex applications. If any organization wants to evolve its technology stack, microservices help in achieving that.

Microservice architecture is not going to solve all the problems of the world. In fact, like everything, it has several disadvantages, complexities, and problems. When using this architecture, numerous problems emerge and, of course, you must solve them all.

 Applying microservices in the wrong place or at the wrong time could generate more problems than improvements.

In addition, it is necessary to change paradigms to define and create microservices, and that does not involve simply breaking a large solution into smaller pieces.

In summary, the microservice architecture style is an approach that creates a single application organized by a suite of small services. Another interesting aspect is that each service could run independently, meaning that there are no restrictions on the technology. As mentioned before, these services are built based on **business capabilities.**

Another important piece of information is that they should work through fully automated and independent deployment mechanisms, which means that if one microservice breaks, the other ones should still go to production if they are green.

The first time this term appeared was on Martin Fowler's website (https://martinfowler.com/articles/microservices.html). Here is a visual representation:

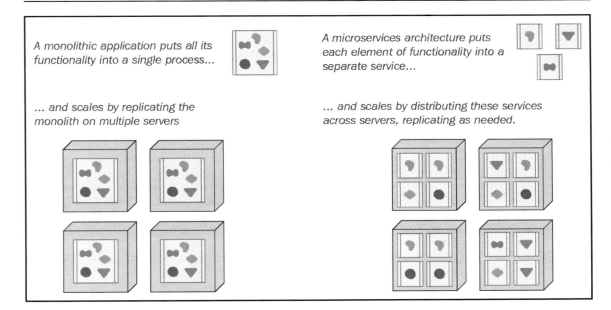

Differences between monolithic applications and microservices. Source https://martinfowler.com/articles/microservices.html

It is highly recommended that you read the full original article from Martin Fowler and James Lewis at `https://martinfowler.com/articles/microservices.html`.

Breaking down order-api

Having introduced microservices, we are going to split our application into two microservices so that they can be deployed, changed, and scaled independently.

Basically, our application was created with a lot of principles used by microservices, such as being partially organized around business capabilities, smart endpoints, infrastructure automation, designing for failure, and so on.

The idea now is to create two applications, each one with its own database, GitHub repository, and CI/CD, and the endpoints will be exposed by Google Cloud Endpoints:

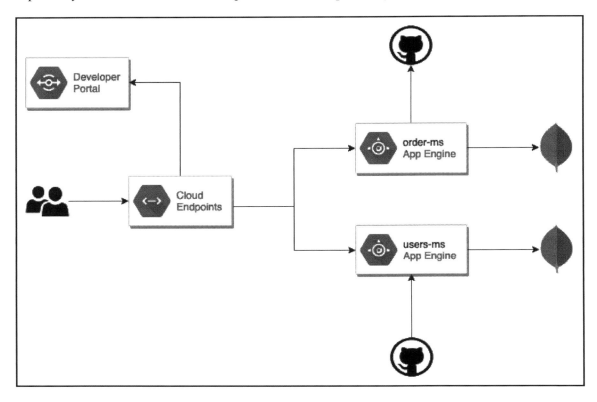

Breaking down order-api into two microservices

Each microservice will be responsible only for its responsibility, such as the order microservices doesn't need to handle login either user microservice. Since we are going to use Google Cloud Endpoints, we can let endpoints handle that for us with an API key.

In order to create this strategy, we will follow these steps in the next sections:

1. Create databases for each service
2. Create a GitHub account for order and user microservices
3. Design the Swagger for those services
4. Implement the services, and run and test them locally
5. Create projects on Google Cloud for the services

6. Enable the services and APIs needed for deployment
7. Create a Travis pipeline for both services
8. Create an API key on Google Cloud
9. Apply Continuous Delivery practices

Creating the databases

Before we get started with the application itself, we will first create the databases for those services at mLab, the same way we did for `order-api`.

- The first one is called `order-ms`:

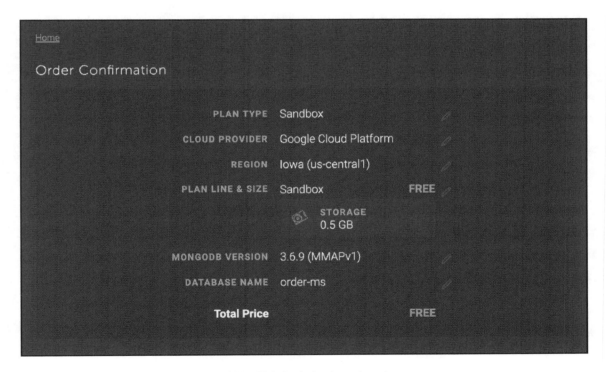

mLab MongoDB database for the order-ms microservice

- The other one is called `user-ms`:

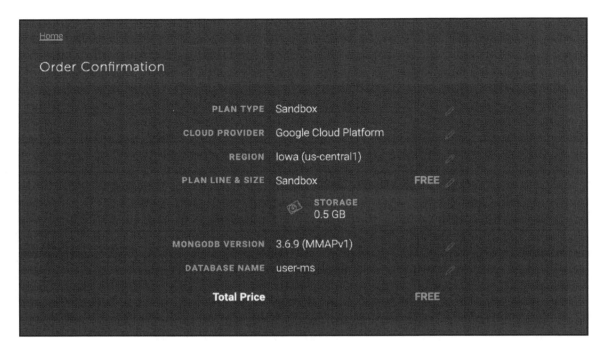

mLab MongoDB database for the user-ms microservice

Also, create users for those databases:

```
user: order-ms-user
pass: 83kerjIkdf
```

Creating a password for `user-ms-user`:

```
user: user-ms-user
pass: kdIdf934nK
```

 Feel free to choose any password you want.

Test the database connection using Robomongo and move on to the next section.

Creating the projects on GitHub

Go to your GitHub account and create two new projects there for each one of the services.

- For `user-ms`, this should look as follows:

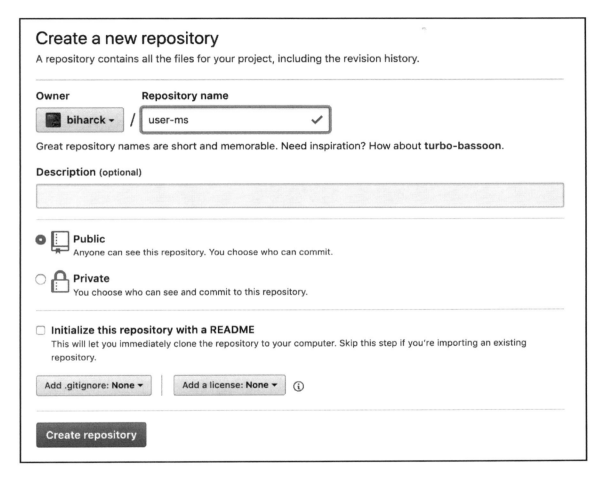

The user microservice GitHub repository

- For `order-ms`, your screen should look like the following screenshot:

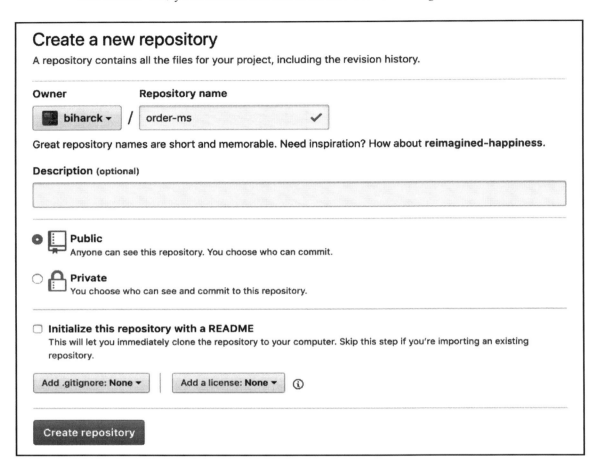

The order microservice GitHub repository

Clone the `user-ms.git` repository locally, as follows:

```
$ git clone git@github.com:<YOUR_GITHUB_USER>/user-ms.git
```

Clone the `order-ms.git` locally:

```
$ git clone git@github.com:<YOUR_GITHUB_USER>/order-ms.git
```

GCP projects

Following the same idea as in the previous chapter, go to the Google Cloud Platform console and create two new projects there—one called `order-ms` and another one called `users-ms`. After you create those two projects, grab the project ID for both of them and keep them in a separate place because we will need this information in the next section:

Google Cloud projects view

Swagger implementation

Basically, we will split the Swagger we created in `Chapter 3`, *Designing RESTful APIs with OpenAPI and Swagger*, into two new ones with fewer operations, since there is no need to log in anymore.

At the time of writing, Google Cloud Endpoints supports only Swagger Version 2: `https://cloud.google.com/endpoints/docs/openapi/`.

The `order-ms` Swagger will be as follows:

```
# [START swagger]
swagger: '2.0'
info:
  description: Order Microservice
  title: Order Microservice
  version: 1.0.0
host: order-ms-227100.appspot.com
# [END swagger]
consumes:
  - application/json
```

```
produces:
  - application/json
  - application/xml
schemes:
  - https
  - http
paths:
  /api:
    get:
      tags:
        - api
      description: Returns a default api message
      operationId: getApi
      security:
        - api_key: []
      produces:
        - application/json
      responses:
        200:
          description: successful operation
  /store/inventory:
    get:
      tags:
        - store
      summary: Returns user inventories from the store
      description: Returns a map of status codes to quantities
      operationId: getInventory
      responses:
        200:
          description: successful operation
      security:
        - api_key: []
  /store/orders:
    post:
      tags:
        - store
      summary: Place an order for a user
      operationId: addOrder
      produces:
        - 'application/json'
      responses:
        201:
          description: successful operation
          schema:
            $ref: '#/definitions/Order'
        400:
          description: Invalid Order
      security:
```

```
      - api_key: []
    parameters:
      - description: Order information
        in: body
        name: message
        required: true
        schema:
          $ref: '#/definitions/Order'
/store/orders/{orderId}:
  get:
    tags:
      - store
    summary: Find purchase order by ID
    operationId: getOrder
    produces:
      - 'application/json'
      - 'application/xml'
    responses:
      201:
        description: successful operation
        schema:
          $ref: '#/definitions/Order'
      400:
        description: Invalid ID supplied
      404:
        description: Order not found
    security:
      - api_key: []
    parameters:
      - in: path
        name: orderId
        type: integer
        required: true
        description: Numeric ID of the order to get.
  delete:
    tags:
      - store
    summary: Deletes the order with the specified ID.
    operationId: removeOrder
    responses:
      204:
        description: User was deleted.
      400:
        description: Invalid ID supplied
      404:
        description: Order not found
    security:
      - api_key: []
```

```
        parameters:
          - in: path
            name: orderId
            type: integer
            required: true
            description: Numeric ID of the order to get.
  definitions:
    Order:
      type: object
      properties:
        id:
          type: integer
          format: int64
        userId:
          type: integer
          format: int64
        quantity:
          type: integer
          format: int32
        shipDate:
          type: string
          format: date-time
        status:
          type: string
          description: Order Status
          enum:
            - placed
            - approved
            - delivered
        complete:
          type: boolean
          default: false
      xml:
        name: Order
  # This section requires all requests to any path to require an API key.
  security:
    - api_key: []
  securityDefinitions:
    api_key:
      type: 'apiKey'
      name: 'key'
      in: 'query'
```

Pay attention to the host parameter, since you have to specify the host based on the project ID you created for order-ms:

```
host: order-ms-227100.appspot.com
```

The following screenshot shows Swagger visualization of **Order Microservice**:

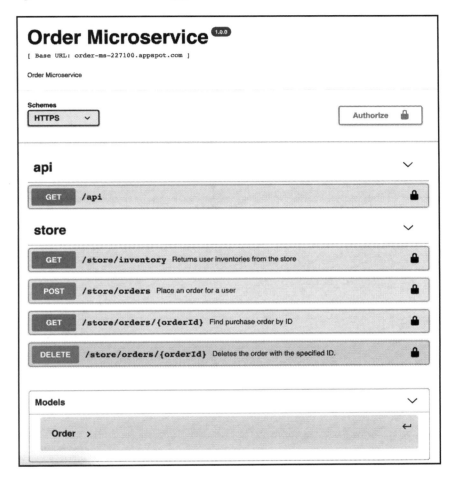

The rder microservice Swagger visualization

We will do the same thing with the user-ms swagger:

```
# [START swagger]
swagger: '2.0'
info:
  description: User Microservice
  title: User Microservice
  version: 1.0.0
host: user-ms-227100.appspot.com
# [END swagger]
```

```
consumes:
  - application/json
produces:
  - application/json
  - application/xml
schemes:
  - https
  - http
paths:
  /api:
    get:
      tags:
        - api
      description: Returns a default api message
      operationId: getApi
      security:
        - api_key: []
      produces:
        - application/json
      responses:
        200:
          description: successful operation
  /users:
    post:
      tags:
        - store
      summary: Create a user
      operationId: addUser
      produces:
        - 'application/json'
      responses:
        201:
          description: successful operation
          schema:
            $ref: '#/definitions/User'
        400:
          description: Invalid User
      security:
        - api_key: []
      parameters:
        - description: User information
          in: body
          name: message
          required: true
          schema:
            $ref: '#/definitions/User'
  /users/{username}:
    get:
```

```
    tags:
      - store
    summary: Find user by username
    operationId: getUser
    produces:
      - 'application/json'
      - 'application/xml'
    responses:
      201:
        description: successful operation
        schema:
          $ref: '#/definitions/User'
      400:
        description: Invalid ID supplied
      404:
        description: User not found
    security:
      - api_key: []
    parameters:
      - in: path
        name: username
        type: string
        required: true
        description: username of the user to get.
  patch:
    tags:
      - store
    summary: Update the user with the specified username.
    operationId: updateUser
    responses:
      204:
        description: User was updated.
      400:
        description: Invalid ID supplied
      404:
        description: User not found
    security:
      - api_key: []
    parameters:
      - in: path
        name: username
        type: string
        required: true
        description: username of the user to update.
      - in: body
        name: message
        required: true
        schema:
```

```
                $ref: '#/definitions/User'
    delete:
      tags:
        - store
      summary: Deletes the user with the specified username.
      operationId: removeUser
      responses:
        204:
          description: User was deleted.
        400:
          description: Invalid ID supplied
        404:
          description: User not found
      security:
        - api_key: []
      parameters:
        - in: path
          name: username
          type: string
          required: true
          description: username of the user to delete.
definitions:
  User:
    type: object
    properties:
      _id:
        type: integer
        format: int64
      username:
        type: string
      firstName:
        type: string
      lastName:
        type: string
      email:
        type: string
      password:
        type: string
      phone:
        type: string
      userStatus:
        type: integer
        format: int32
        description: User Status
    xml:
      name: User
# This section requires all requests to any path to require an API key.
security:
```

```
 - api_key: []
securityDefinitions:
  api_key:
    type: 'apiKey'
    name: 'key'
    in: 'query'
```

As with `order-ms`, be careful with the host information for `user-ms`:

host: user-ms-227100.appspot.com

The following screenshot shows Swagger visualization of **User Microservice**:

The user microservice Swagger visualization

As was mentioned before, it is not necessary to keep the operations neither login or logout since the API key is going to handle that for us from Google Cloud Endpoints.

Create the skeleton for both microservices

Summarizing, so far, we have accomplished the following:

- Created databases for each service
- Created GitHub accounts for the order and user microservices
- Designed the Swagger for those services
- Created projects on Google Cloud for them

So, the next step will be implementing the services. The next sections will help you with that. In general, we will grab almost the piece of code we generated in the previous chapters and split them into two microservices due to the domain being already well defined.

The order-ms code

The `order-ms` structure has to be everything from the `order-api` service we created without users routes. So, that being said, the structure will be as follows:

The order-ms project structure

Now, we have to create a service account for the `order-ms` microservice on GCP, download the key, and generate a secure version of the key for Travis:

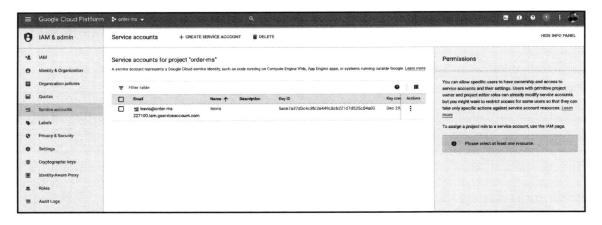

Service account for order-ms

 When you download the key, move it to the `order-ms` project and rename it `gce.json`.

To generate a secure version for Travis CI, run the following command:

```
$ travis encrypt-file gce.json --add
```

Then, a line similar to the one that follows here will appear in your `.travis.yml` file:

```
before_install:
- openssl aes-256-cbc -K $encrypted_39e82dbbfc3e_key -iv
$encrypted_39e82dbbfc3e_iv
  -in gce.json.enc -out gce.json -d
```

Now, it's time to add the environment variables to Google Cloud Storage. Create a new bucket with the same name as your project ID, `order-ms-227100`, for instance, as shown here:

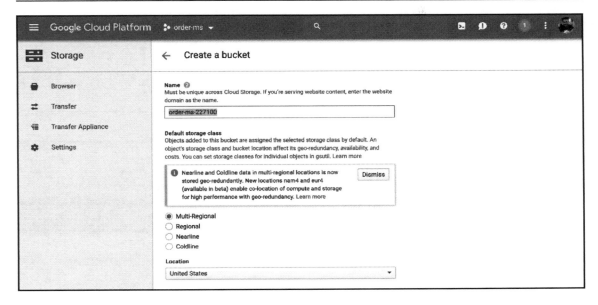

Google Cloud Storage for the order-ms service

Then, upload a file there called `.env.prod` with the content you got from mLab:

```
MONGODB_URL_PORT=ds125381.mlab.com:25381
MONGODB_DATABASE=order-ms
MONGODB_USER=order-ms-user
MONGODB_PASS=83kerjIkdf
```

The following screenshot shows the **Bucket details** in Google Cloud Storage:

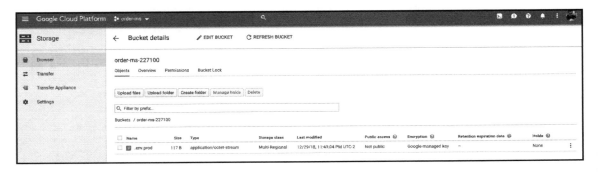

The .env.prod file on Google Cloud Storage

Since we have everything in place, run the tests to make sure everything is fine:

```
$ npm install
```

The following, is the command to run `test`:

```
$ npm run test
baseRoute
    should respond with HTTP 200 status (63ms)
    should respond with success message

  orderRoute
[2018-12-30T02:11:14.619Z] [info] => [GET] [/store/orders/] 000
[2018-12-30T02:11:14.632Z] [info] => [GET] [/store/orders/:{orderId}] Order
000 not found.
    should respond with HTTP 404 status because there is no order
[2018-12-30T02:11:14.663Z] [info] => [POST] [/store/orders/] 20
[2018-12-30T02:11:14.669Z] [info] => [POST] [/store/orders/] { _id:
5c28294258e64380ee8a0b8b,
  userId: '20',
  quantity: 1,
  shipDate: 2018-12-30T02:11:14.514Z,
  status: 'PLACED',
  complete: false }
    should create a new order and retrieve it back (78ms)
[2018-12-30T02:11:14.716Z] [info] => [GET] [/store/orders/]
5c28294258e64380ee8a0b8b
    should return the order created on the step before
[2018-12-30T02:11:14.728Z] [info] => [GET] [/store/orders/]
    should return all orders so far
[2018-12-30T02:11:14.738Z] [info] => [GET] [/store/orders/]
    should not return orders because offset is higher than the size of the
orders array
[2018-12-30T02:11:14.745Z] [info] => [GET] [/store/inventory/] PLACED
    should return the inventory for all users
[2018-12-30T02:11:14.755Z] [warn] => [DELETE] [/store/orders/]
5c28294258e64380ee8a0b8b
    should remove an existing order
[2018-12-30T02:11:14.763Z] [warn] => [DELETE] [/store/orders/]
5c28294258e64380ee8a0b8b
[2018-12-30T02:11:14.768Z] [warn] => [DELETE] [/store/orders/:{orderId}]
Order id 5c28294258e64380ee8a0b8b not found
    should return 404 when it is trying to remove an order because the
order does not exist


    10 passing (255ms)
```

Now run the following command:

```
$ npm run test:mutation
```

The following screenshot shows the **Mutation score**:

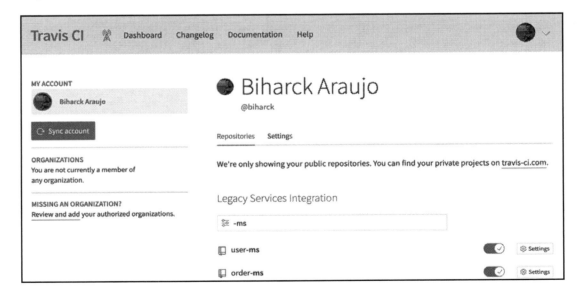

File / Directory	Mutation score		# Killed	# Survived	# Timeout	# No coverage	# Runtime errors	# Transpile errors	Total detected	Total undetected	Total mutants
All files	100.00 %	100.00	3	0	3	0	144	0	6	0	150
controllers/	100.00 %	100.00	1	0	1	0	65	0	2	0	67
models/	100.00 %	100.00	0	0	0	0	5	0	0	0	5
routes/	100.00 %	100.00	0	0	0	0	8	0	0	0	8
utility/	100.00 %	100.00	0	0	0	0	29	0	0	0	29
app.ts	100.00 %	100.00	1	0	2	0	28	0	3	0	31
schemas/order.ts	100.00 %	100.00	1	0	0	0	6	0	1	0	7
server.ts	100.00 %	100.00	0	0	0	0	3	0	0	0	3

Generated with stryker-html-reporter generator. Visit the Stryker website

Mutation tests for order-ms

Now, go to the Travis CI **Dashboard** and activate the `order-ms` service; take advantage of being there, and enable `user-ms` as well:

Enabling Travis CI for both applications

Apply the first deployment manually, as we did before for `order-ms`. First, you need to change the account with the following command:

```
$ gcloud init
```

This command will show the following output:

```
Welcome! This command will take you through the configuration of gcloud.

Settings from your current configuration [typescript3] are:
compute:
  region: us-east1
  zone: us-east1-b
core:
  account: baraujo@thoughtworks.com
  disable_usage_reporting: 'True'
  project: order-ms

Pick configuration to use:
 [1] Re-initialize this configuration [typescript3] with new settings
 [2] Create a new configuration
 [3] Switch to and re-initialize existing configuration: [default]
Please enter your numeric choice: 3
```

Then, log into your account, which, in most cases, is option 1:

```
Your current configuration has been set to: [default]

You can skip diagnostics next time by using the following flag:
  gcloud init --skip-diagnostics

Network diagnostic detects and fixes local network connection issues.
Checking network connection...done.
Reachability Check passed.
Network diagnostic passed (1/1 checks passed).

Choose the account you would like to use to perform operations for
this configuration:
 [1] baraujo@...
 [2] Log in with a new account
```

Then, select the `order-ms` project.

Before we deploy the application, change the `app.yaml` file to deploy in the same way we did before:

```
runtime: custom
env: flex
```

```
manual_scaling:
  instances: 1
resources:
  cpu: 1
  memory_gb: 0.5
  disk_size_gb: 10
# endpoints_api_service:
# #remeber to replace the name with your project id
# name: order-ms-227100.appspot.com
# rollout_strategy: managed
# liveness_check:
# path: '/api'
# check_interval_sec: 30
# timeout_sec: 4
# failure_threshold: 2
# success_threshold: 2
```

Run the `deploy` command and wait for it to complete:

```
$ gcloud app deploy
```

Once that's done, try to reach the /api endpoint there using POSTman or any other tool you prefer. You should see the following output:

```
{
    "title": "Order microservice"
}
```

Now, let's go back and take a look at the `app.yaml` file with the uncommented lines:

```
runtime: custom
env: flex
manual_scaling:
  instances: 1
resources:
  cpu: 1
  memory_gb: 0.5
  disk_size_gb: 10
endpoints_api_service:
  #remeber to replace the name with your project id
  name: order-ms-227100.appspot.com
  rollout_strategy: managed
liveness_check:
  path: '/api'
  check_interval_sec: 30
  timeout_sec: 4
  failure_threshold: 2
  success_threshold: 2
```

There are two new main items there—`endpoints_api_service`, which is related to Google Cloud Endpoints, and `liveness_check`, which is related to the health check. We are now going to deploy the Swagger specification and link the contract with the microservice.

 Remember to replace `endpoints_api_service` | `name with your project id like` `<YOR_PROJECT_ID>.appspot.com`.

Once you replace the project ID, run the following command to deploy the Swagger file:

```
$ gcloud endpoints services deploy openapi-appengine.yaml
```

When this command finishes, go to the Google Cloud console, search for **Cloud Endpoints** and click on **GO TO CLOUD ENDPOINTS**:

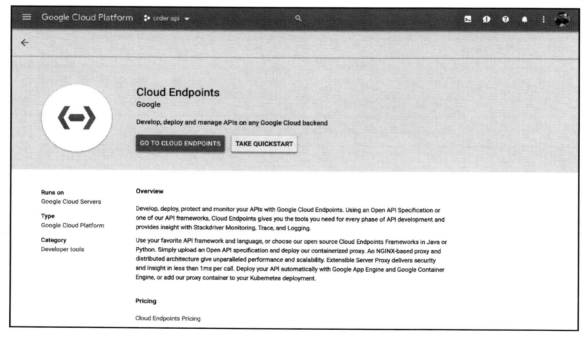

Google Cloud Endpoints

Then, click on **Developer Portal** and **Create portal**:

Activating Developer Portal

This process takes a while, but, when it finishes, you will be able to see the `order-ms` developer portal:

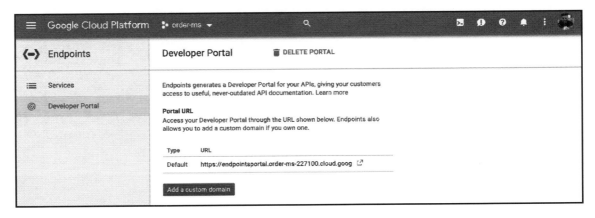

The API Portal created and the URL available

The portal looks as follows:

The order-ms Developer Portal

Here is the portal's main page:

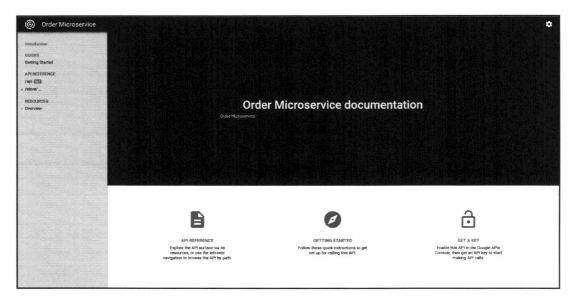

The main page of order-ms on Developer Portal

The `GET /store/orders/{orderID}` definition on **Developer Portal** is shown as follows:

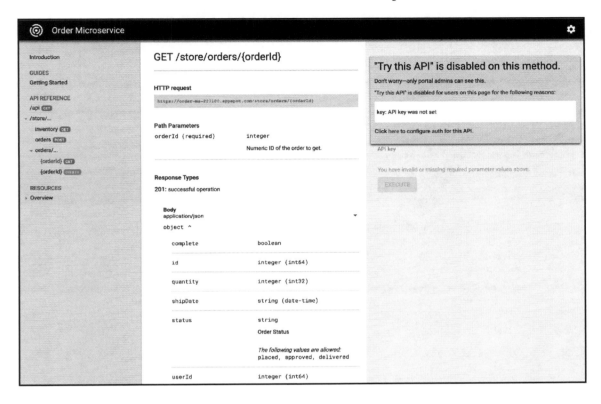

The GET /store/orders/{orderID} definition on Developer Portal

Now, deploy the application again with the changes in the `app.yaml` file:

```
$ gcloud app deploy
```

After that, you will no longer be allowed to perform the request without an API key, which we will create right now. Go to the Google Cloud console and type `API Key`:

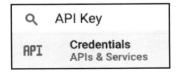

Searching for the API key

Once you're there, click on **Create credentials** | **API Key**:

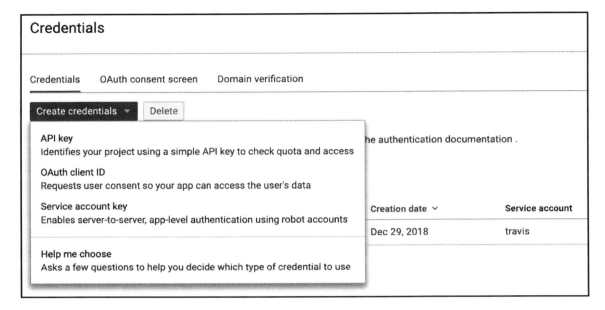

Creating an API key for order-ms

Copy the generated key and include the following information in any requests:

```
https://order-ms-227100.appspot.com/api?key=<YOUR_KEY>
```

Now, you are allowed to perform requests successfully:

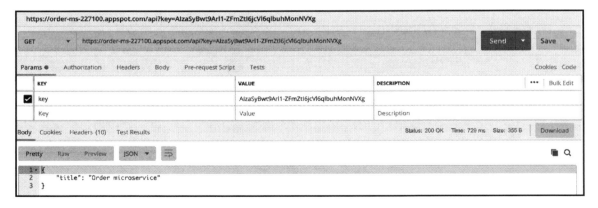

Request from POSTman using the API key

Enable the **App Engine Admin API**:

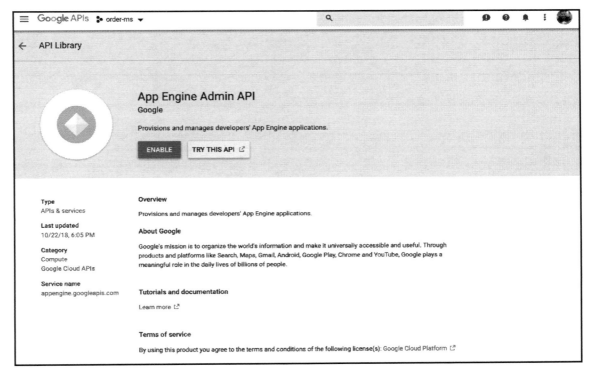

The API library

Finally, commit and push the files to GitHub, and watch Travis CI do the job of deploying the new version:

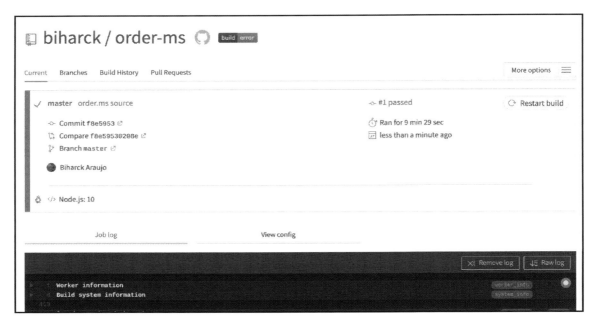

New version deployed by Travis CI as per Continuous Delivery practice

The user-ms code

The next step is to do the same thing with the `user-ms` microservice. Since we are using Google Endpoints, the only reason to have this microservice is if you treat the user as a customer, for example, and maybe you could use the basic authentication with the user password and login as part of the authentication/authorization process.

As you might have guessed, the project structure is almost the same as for order-ms. The user-ms code will have the following structure:

Before we enter the code for user-ms, copy and paste the following files from the order-ms service into the user-ms service:

- tslint.json
- tsconfig.json
- stryker.conf.js
- Dockerfile
- .prettierrc
- .gitignore
- .dockerignore
- test/routes/api.spec/ts
- test.mocha.opts
- src/server.ts
- src/utility/errorHandler.ts
- src/utility/logger.ts
- src/routes/api.ts
- src/models/applicationType.ts

Run the tests, as follows:

```
$ npm run test
    should respond with HTTP 200 status (98ms)
    should respond with success message

  userRoute
[2018-12-30T03:34:43.807Z] [info] => [GET] [/users] NO_USER
[2018-12-30T03:34:43.829Z] [info] => [GET] [/users/:{username}] user with
username NO_USER not found
    should respond with HTTP 404 status because there is no user
[2018-12-30T03:34:43.905Z] [info] => [POST] [/users] { _id: null,
  username: 'John',
  firstName: 'John',
  lastName: 'Doe',
  email: 'John@memail.com',
  password: 'password',
  phone: '5555555',
  userStatus: 1 }
[2018-12-30T03:34:44.021Z] [info] => John
    should create a new user and retrieve it back (189ms)
```

```
[2018-12-30T03:34:44.022Z] [info] => [GET] [/users] John
    should return the user created on the step before
[2018-12-30T03:34:44.030Z] [info] => [PATCH] [/users] John
    should updated the user John
[2018-12-30T03:34:44.046Z] [info] => [GET] [/users] John_Updated
    should return the user updated on the step before
[2018-12-30T03:34:44.055Z] [info] => [PATCH] [/users] Mary
[2018-12-30T03:34:44.058Z] [info] => [PATCH] [/users/:{username}] user with
username Mary not found
    should return 404 because the user does not exist
[2018-12-30T03:34:44.060Z] [warn] => [DELETE] [/users] John_Updated
    should remove an existent user
[2018-12-30T03:34:44.068Z] [warn] => [DELETE] [/users] Mary
[2018-12-30T03:34:44.070Z] [info] => [DELETE] [/users/:{username}] user
with username Mary not found
    should return 404 when it is trying to remove an user because the user
does not exist


    10 passing (401ms)
```

Following, is the command to run `test`:

```
$ npm run test:mutation
```

The result is shown in the following screenshot:

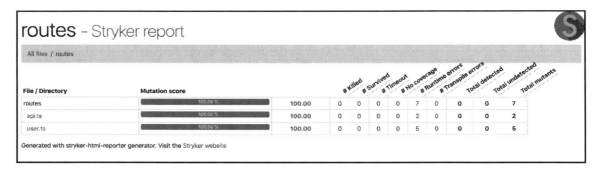

File / Directory	Mutation score		# Killed	# Survived	# Timeout	# No coverage	# Runtime errors	# Transpile errors	Total detected	Total undetected	Total mutants
routes	100.00 %	100.00	0	0	0	0	7	0	0	0	7
api.ts	100.00 %	100.00	0	0	0	0	2	0	0	0	2
user.ts	100.00 %	100.00	0	0	0	0	5	0	0	0	5

Generated with stryker-html-reporter generator. Visit the Stryker website

Stryker output for user routes

After that, redo the same steps as you did for `order-ms` and, in the end, you will have a Travis CI pipeline deploy the application for you, and also a `user-ms` **Developer Portal**:

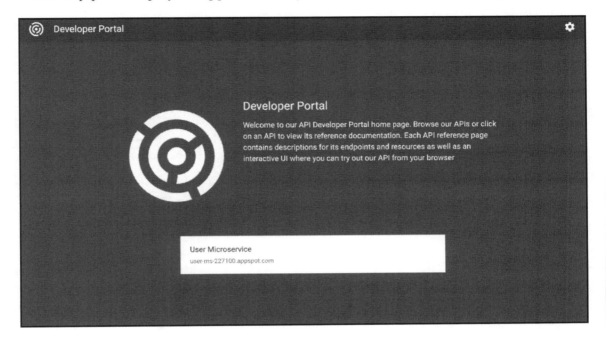

The user-ms Developer Portal home page

The following screenshot shows the **POST/users** definitions:

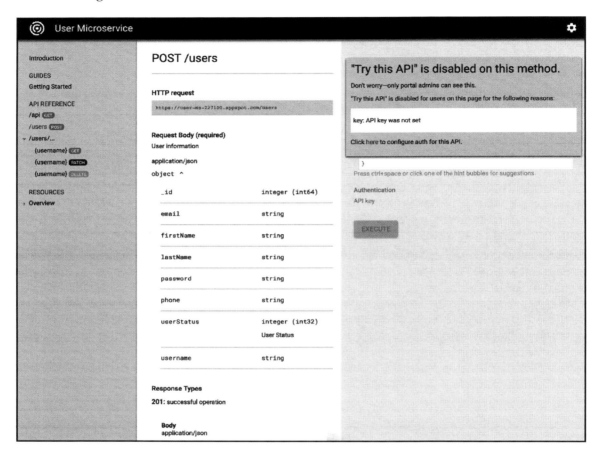

The user-ms POST /user definitions on Developer Portal

The following screenshot shows the new version of `user-ms` to production:

Travis CI promoting a new version of user-ms to production

Summary

This chapter walked you through some best practices when creating microservices. You learned how to split a service that could be scaled up and down separately on demand.

You also learned how to create a minimal microservice ecosystem, which allowed Continuous Delivery. Also, with the isolation of components, you learned that if one API is not working, the rest of the APIs will stay up and running.

The next and final chapter will show you some techniques in relation to GraphQL.

Questions

1. Are microservice applications easier to maintain than monolithic applications?
2. Name at least two advantages of using microservice architecture.
3. What is the relationship between a database and a microservice?
4. Is Continuous Integration part of microservice architecture?
5. Is Continuous Delivery part of microservice architecture?
6. With microservices, is it possible to write services in different languages?
7. What is a **Developer Portal**?

Further reading

In order to improve your knowledge about formatting strategy, the following books are recommended, as they will be helpful in the coming chapters:

- *Modern JavaScript Web Development Cookbook* (`https://www.packtpub.com/web-development/modern-javascript-web-development-cookbook`)
- *Hands-On Microservices with Node.js* (`https://www.packtpub.com/web-development/hands-microservices-nodejs`)

13
Flexible APIs with GraphQL

GraphQL is a new approach to serving data. Some people even define GraphQL as REST 2.0. In this chapter, we will take a look at the differences and similarities between GraphQL and REST. You will also learn how to add support for GraphQL to your existing RESTful API. With some examples of querying data, as well as validating and executing a query, you will learn how to benefit the most from using GraphQL with your RESTful APIs.

The following topics will be covered in this chapter:

- Introduction to GraphQL
- Adding support for GraphQL to your REST API
- Queries
- Schemas and types
- Execution

Technical requirements

All of the information that's required to run the code in this chapter can be found in the relevant sections of this chapter. The only requirement is that you have applications such as Node.js, VS Code and TypeScript installed on your system, which we covered in Chapter 4, *Setting Up Your Development Environment*.

All of the code that's used in this chapter is available at `https://github.com/PacktPublishing/Hands-On-RESTful-Web-Services-with-TypeScript-3/tree/master/Chapter13`.

Introduction to GraphQL

As we mentioned at the beginning of this chapter, GraphQL is an interesting technique that provides a query language for APIs. We can apply this to existing data to fulfil any queries that are requested. A key factor of GraphQL is that it provides an interesting and understandable description of the data that comes from an API. This strategy enables clients to ask for exactly what they need and nothing more. So, if, as a requester, I just want two fields, why should I receive a ton of unnecessary information? GraphQL makes the API easier to adapt to the current situation.

 It is really recommended that you take a look at the GraphQL web page to learn more: `https://graphql.org/`.

In summary, a GraphQL service is composed of the following:

- Defining types and fields on those types
- Providing functions for each field on each type

Taking the example from the GraphQL website, suppose that a GraphQL service tells us who the logged-in user is and the user's name. Based on this example, the types should be as follows:

```
type Query {
  me: User
}

type User {
  id: ID
  name: String
}
```

The functions for each field on each type should be as follows:

```
function Query_me(request) {
  return request.auth.user;
}

function User_name(user) {
  return user.getName();
}
```

Let's say we provide the following query:

```
{
  me {
    name
  }
}
```

The response might look something like this:

```
{
  "me": {
    "name": "Luke Skywalker"
  }
}
```

 Before we get started with changes, we will reinforce that it is strongly recommended that you study the *Learn* section on GraphQL's website: https://graphql.org/learn/queries/.

Configuring GraphQL with the order-ms service

Let's take the `order-ms` service as an example and set up GraphQL there. First, we will need to install some dependencies:

```
$ npm install --save apollo-server-express graphql-tools graphql
```

Then, we need to create a new file called `graphql.ts` under `src/graphql`, with the following content:

```
import { ApolloServer, gql } from 'apollo-server-express'

export class GraphQL {
  public typeDefs: string
  public resolvers: Object
  public server: ApolloServer

  constructor() {
    this.typeDefs = gql`
      type Query {
        hello: String
      }
    `

    this.resolvers = {
```

```
      Query: {
        hello: () => 'Hello world!',
      },
    }

    this.server = new ApolloServer({
      typeDefs: this.typeDefs,
      resolvers: this.resolvers,
    })
  }

  public setup(app): void {
    this.server.applyMiddleware({ app: app })
  }
}
```

Basically, we are creating a *Hello World* GraphQL. Under this file, we have a class that has a constructor() where it defines the types and resolvers:

```
constructor() {
    this.typeDefs = gql`
      type Query {
        hello: String
      }
    `

    this.resolvers = {
      Query: {
        hello: () => 'Hello world!',
      },
    }

    this.server = new ApolloServer({
      typeDefs: this.typeDefs,
      resolvers: this.resolvers,
    })
  }
```

By definition, types define what is expected and resolvers explain what data we are getting. The second part of this class is a function that applies the middleware to the express application:

```
public setup(app): void {
    this.server.applyMiddleware({ app: app })
  }
```

After you create this file, go to the `src/app.ts` file and make the following changes to allow the `order-ms` application to use GraphQL:

```
import * as bodyParser from 'body-parser'
import * as dotenv from 'dotenv'
import * as express from 'express'
import * as expressWinston from 'express-winston'
import * as mongoose from 'mongoose'
import * as winston from 'winston'
import { GraphQL } from './routes/graphql'
import { APIRoute } from './routes/api'
import { OrderRoute } from './routes/order'
import * as errorHandler from './utility/errorHandler'
import { OrderAPILogger } from './utility/logger'

class App {
  public app: express.Application
  public apiRoutes: APIRoute = new APIRoute()
  public orderRoutes: OrderRoute = new OrderRoute()
  public graphQL: GraphQL = new GraphQL()
  public mongoUrl: string
  public mongoUser: string
  public mongoPass: string

  constructor() {
    const path = `${__dirname}/../.env.${process.env.NODE_ENV}`

    dotenv.config({ path: path })
    this.mongoUrl = `mongodb://${process.env.MONGODB_URL_PORT}/${
      process.env.MONGODB_DATABASE
    }`
    this.mongoUser = `${process.env.MONGODB_USER}`
    this.mongoPass = `${process.env.MONGODB_PASS}`

    this.app = express()
    this.graphQL.setup(this.app)

    OrderAPILogger.logger.info(
      `graphql running at ${this.graphQL.server.graphqlPath}`
    )

    this.app.use(bodyParser.json())
    this.apiRoutes.routes(this.app)
    this.orderRoutes.routes(this.app)
    this.mongoSetup()
    this.app.use(
      expressWinston.errorLogger({
        transports: [new winston.transports.Console()],
```

```
      })
    )
    this.app.use(errorHandler.logging)
    this.app.use(errorHandler.clientErrorHandler)
    this.app.use(errorHandler.errorHandler)
  }

  private mongoSetup(): void {
    let options

    if (process.env.NODE_ENV !== 'prod') {
      options = {
        useNewUrlParser: true,
      }
    } else {
      options = {
        user: this.mongoUser,
        pass: this.mongoPass,
        useNewUrlParser: true,
      }
    }
    mongoose.connect(
      this.mongoUrl,
      options
    )
  }
}

export default new App().app
```

Then, start the application:

```
$ npm run dev
```

Now, go to your browser and open the following URL:

```
http://localhost:3000/graphql
```

It should open a GraphQL playground so that you can test your changes:

GraphQL playground web page

Let's move on and try the first query, which is based on the data that was provided by the resolver:

```
Query: {
    hello: () => 'Hello world!',
}
```

Try the query:

```
query{
  hello
}
```

The response should be something like this:

```
{
  "data": {
    "hello": "Hello world!"
  }
}
```

The following screenshot shows the response on GraphQL:

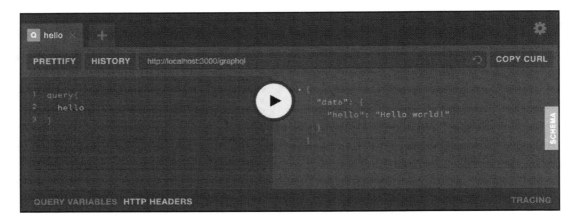

First query with the template data

Improving order-ms with order data

Let's improve on `order-ms` a little bit. Here, we will treat some order data and move the `types` and `resolvers` from the `graphql.ts` file to their specific files. To do that, create a file called `types.ts` with the following content:

```ts
import { gql } from 'apollo-server-express'
import { DocumentNode } from 'graphql'

export class OrderGraphQLTypes {
  public getTypes(): DocumentNode {
    return gql`
      type Order {
        id: ID!
        userId: Int!
        quantity: Int!
        status: String!
        complete: Boolean!
      }
      type Query {
        allOrders: [Order]!
        listByOrderId(id: ID): Order!
      }

    `
  }
}
```

Note that two types are now defined:

- **Order**: The order itself
- **Query**: Defines the `allOrders` and `listByOrderId` functions

Now, create a new file called `resolvers.ts` with the following content:

```
import { OrderStatus } from '../models/orderStatus'

const orders = [
  {
    id: 1,
    userId: '1',
    quantity: 2,
    status: OrderStatus.Placed,
    complete: false,
  },
  {
    id: 2,
    userId: '2',
    quantity: 1,
    status: OrderStatus.Placed,
    complete: false,
  },
  {
    id: 3,
    userId: '3',
    quantity: 1,
    status: OrderStatus.Approved,
    complete: false,
  },
  {
    id: 4,
    userId: '1',
    quantity: 10,
    status: OrderStatus.Delivered,
    complete: true,
  },
]

export class OrderGraphQLResolvers {
  public getResolvers(): IResolvers {
    return {
      Query: {
        allOrders: () => orders,
        listByOrderId: (root, args, context) => {
          return orders.find(order => order.id === Number(args.id))
```

```
          },
        },
      }
    }
  }
```

We created a fake array of orders for this example and a class called `OrderGraphQLResolvers`, which is going to handle the `allOrders` and `listByOrderId` functions for us.

Now, go back to the `graphql.ts` file and change the way we are setting the `types` and `resolvers` so that we can use those new classes:

```typescript
import { ApolloServer, IResolvers } from 'apollo-server-express'
import { DocumentNode } from 'graphql'
import { OrderGraphQLResolvers } from './resolvers'
import { OrderGraphQLTypes } from './types'

export class GraphQL {
  public typeDefs: DocumentNode
  public resolvers: IResolvers
  public server: ApolloServer

  constructor() {
    this.typeDefs = new OrderGraphQLTypes().getTypes()
    this.resolvers = new OrderGraphQLResolvers().getResolvers()

    this.server = new ApolloServer({
      typeDefs: this.typeDefs,
      resolvers: this.resolvers,
    })
  }

  public setup(app): void {
    this.server.applyMiddleware({ app: app })
  }
}
```

Start the application and test it out to get `allOrders`:

Getting allOrders

We also need to get the `order` by ID:

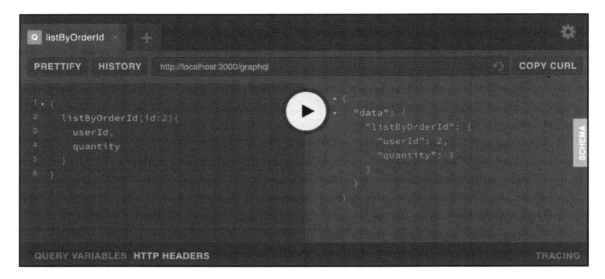

Getting the order by ID

Connecting with MongoDB

Now that we know how to use GraphQL, we can connect with real data through `mongoose`. To do that, we have to change the `resolvers.ts` file to get the data from MongoDB. The changes are really simple—we just have to include the `OrderModel` there, along with its respective `mongoose` operation:

```
import { IResolvers } from 'graphql-tools'
import { OrderModel } from '../schemas/order'

export class OrderGraphQLResolvers {
  public getResolvers(): IResolvers {
    return {
      Query: {
        allOrders: async () => await OrderModel.find({}),
        listByOrderId: async (root, args, context) => {
          return await OrderModel.findById({ _id: args.id })
        },
      },
    }
  }
}
```

After that, restart the server and run the same queries again:

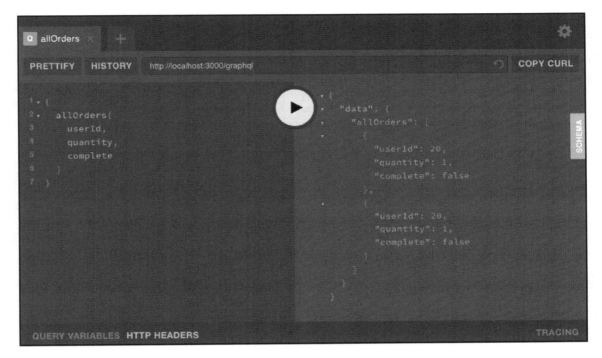

Getting all data from MongoDB

The following screenshot shows the `order` by ID:

Getting an order by ID from MongoDB

Since we are able to get data, we will also expose a way to create a new order called **Mutation**. Briefly, mutation is a way in which you can change/create the state of the data.

Change the `resolvers.ts` file by adding the `createOrder` mutation, as follows:

```
import { IResolvers } from 'graphql-tools'
import { OrderModel } from '../schemas/order'

export class OrderGraphQLResolvers {
  public getResolvers(): IResolvers {
    return {
      Query: {
        allOrders: async () => await OrderModel.find({}),
        listByOrderId: async (root, args, context) => {
          return await OrderModel.findById({ _id: args.id })
        },
      },
      Mutation: {
        createOrder: async (root, args) => {
          const newOrder = new OrderModel(args)
          return await newOrder.save()
        },
      },
    }
  }
}
```

Also, change the types:

```
import { gql } from 'apollo-server-express'

export class OrderGraphQLTypes {
  public getTypes(): DocumentNode {
    return gql`
      type Order {
        id: ID!
        userId: Int!
        quantity: Int!
        status: String!
        complete: Boolean!
      }
      type Query {
        allOrders: [Order]!
        listByOrderId(id: ID): Order!
      }
      type Mutation {
        createOrder(
          userId: Int!
          quantity: Int!
          status: String!
          complete: Boolean!
        ): Order
      }

    }
}
```

Restart the server and try to create a new `order`:

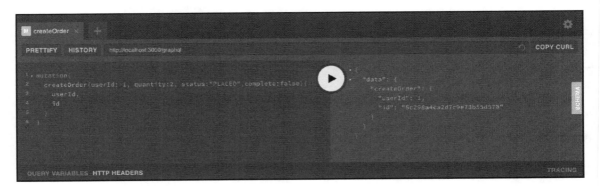

CreateOrder mutation

To make sure that everything is fine, call `allOrders` again. You should see the new resource there:

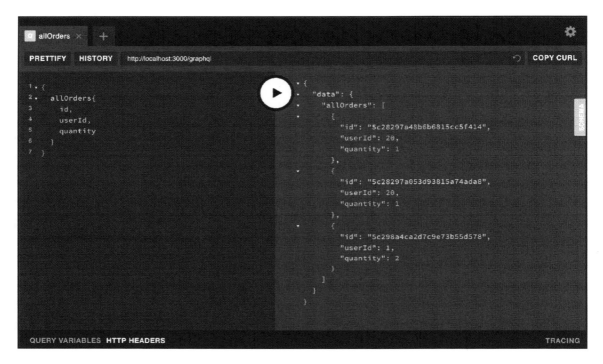

All orders with the new data created by the mutation

Creating a client with Vue.js

Since our microservice is working with GraphQL, we would like to create a simple UI so that we can connect and expose the data we created and interact with it.

Vue.js is a progressive web framework for frontend applications that was created by Evan You. Vue.js stands out because of its simplicity in performing the same tasks as other frameworks.

Since this book is not about Vue.js, we recommend that you go to the Vue.js documentation to learn more: `https://vuejs.org/`.

The first step is to install Vue.js and the Vue.js CLIs:

```
$ npm install -g @vue/cli-init vue
```

After that, create a new project called `client` with the following command:

```
$ vue init webpack client

? Project name client
? Project description Order client
? Author Biharck Araujo <biharck@gmail.com>
? Vue build standalone
? Install vue-router? Yes
? Use ESLint to lint your code? Yes
? Pick an ESLint preset Standard
? Set up unit tests No
? Setup e2e tests with Nightwatch? No
? Should we run `npm install` for you after the project has been created?
(recommended) npm
```

After the installation finishes, go to the `client` folder and start the application:

```
$ npm run dev
```

The application should be available at the following URL:

```
$ http://localhost:8080
```

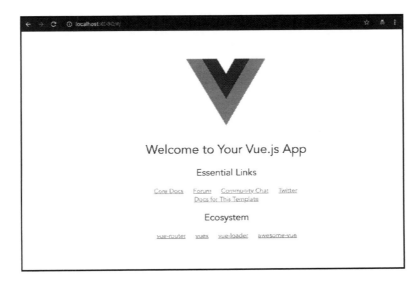

Vue.js default application home page

Now, we will configure the dependencies to connect to order-ms with GraphQL. The following are the dependencies that we will need to configure:

- apollo-client: This is for connecting the client with GraphQL and also managing the state of the data
- vue-apollo: This is responsible for binding the data with the Vue.js application
- graphql-tag: This is for making the query and mutation to the backend

The following command is used to install these libraries:

```
$ npm install --save graphql vue-apollo apollo-client graphql-tag
```

With these libraries in place, we can move on and build the frontend. First, create a folder called src/graphql with a file named allOrders.js. This file will contain the GraphQL query for allOrders:

```
import gql from 'graphql-tag'

export const ALL_ORDERS_QUERY = gql`{
    allOrders{
        id,
        userId,
        quantity
    }
}
`
```

After that, create a new file called src/apolloClient.js:

```
import { createBatchingNetworkInterface, ApolloClient } from 'apollo-client'

export const apolloClient = new ApolloClient({
  networkInterface: createBatchingNetworkInterface({
    uri: 'http://localhost:3000/graphql'
  }),
  connectToDevTools: true
})
```

This file will connect with our backend service using the apollo client library.

Remember to replace the URI with yours if it is different to the default URI.

Create another file, called `src/apolloProvider.js`, with the following content:

```
import Vue from 'vue'
import VueApollo from 'vue-apollo'

import {apolloClient} from './apolloClient'

Vue.use(VueApollo)

export const apolloProvider = new VueApollo({
  defaultClient: apolloClient,
  defaultOptions: {
    $loadingKey: 'loading'
  }
})
```

Now, we are good to write the code to show the orders. Replace the `HelloWorld.vue` file with the following content:

```
<template>
  <div>
    <h4 v-if="loading">Loading......</h4>
    <div v-for="order in allOrders" :key="order.id">
      <router-link :to="order.id" exact>
        <h3>Order:{{order.id}}</h3>
      </router-link>
      <p>
        User: {{order.userId}}
        Quantity:{{order.quantity}}
      </p>
    </div>
  </div>
</template>
<script>
import { ALL_ORDERS_QUERY } from '../graphql/allOrders'
export default {
  name: 'order',
  data () {
    return {
      loading: 0,
      allOrders: []
    }
  },
  apollo: {
    allOrders: {
      query: ALL_ORDERS_QUERY
    }
  }
```

```
}
</script>
<!-- Add "scoped" attribute to limit CSS to this component only -->
<style scoped>
</style>
```

Start the application and open it in a browser of your choice:

Listing orders with the Vue.js application

Now, let's create a new page to show the order by ID. Following the same idea as before, create a new file called `src/graphql/listByOrderId.js` with the following content:

```
import gql from 'graphql-tag'

export const ORDER_BY_ID = gql`
  query orderById($id: ID!){
    listByOrderId(id: $id){
        userId,
        quantity,
        status,
        complete
    }
}`
```

Then, create the `.vue` page file called `src/components/OrderById.vue`:

```
<template>
<div>
<h3>UserID: {{listByOrderId.userId}}</h3>
<p>Quantity: {{listByOrderId.quantity}}</p>
<p>Status: {{listByOrderId.status}}</p>
<p>Complete: {{listByOrderId.complete}}</p>
</div>
</template>
<script>
import {ORDER_BY_ID} from '../graphql/listByOrderId'
export default {
  data () {
    return {
      listByOrderId: {},
      loading: 0
    }
  },
  props: ['id'],
  apollo: {
    listByOrderId: {
      query: ORDER_BY_ID,
      variables () {
        return {
          id: this.id
        }
      }
    }
  }
}
</script>
```

Finally, add the route in `src/router/index.js`:

```
import Vue from 'vue'
import Router from 'vue-router'
import HelloWorld from '@/components/HelloWorld'
import OrderById from '@/components/OrderById'

Vue.use(Router)

export default new Router({
  routes: [
    {
      path: '/',
      name: 'HelloWorld',
      component: HelloWorld
```

```
    },
    {
      path: '/:id',
      name: 'orderById',
      component: OrderById,
      props: true
    }
  ]
})
```

When you click the **Order Id** on the web page, you will be redirected to the **OrderById** page:

OrderById page

The only thing that's missing is the order that we're creating, which we will implement. Create a file called `createOder.js` under `src/graphql`, with the following content:

```
import gql from 'graphql-tag'

export const ADD_ORDER = gql`
  mutation addOrder($userId: Int!, $quantity: Int!, $status: String!,
$complete: Boolean! ) {
    createOrder(userId: $userId, quantity: $quantity, status: $status,
complete: $complete ) {
        userId,
        quantity,
```

```
            status,
            complete
        }
    }
```

Add a component called `AddOrder.vue`:

```
<template>
  <form>
    <label for="userId">userId</label>
    <br>
    <input type="text" name="userId" v-model="userId">
    <br>
    <br>
    <label for="quantity">Quantity</label>
    <br>
    <input type="text" name="quantity" v-model="quantity">
    <br>
    <label for="status">Status</label>
    <br>
    <input type="text" name="status" v-model="status">
    <br>
    <label for="complete">Complete</label>
    <br>
    <input type="text" name="complete" v-model="complete">
    <br>
    <button type="submit" @click="addOrder" >Add</button>
  </form>
</template>
<script>
import { ADD_ORDER } from '../graphql/createOrder'
export default {
  data () {
    return {
      userId: '',
      quantity: '',
      status: '',
      complete: ''
    }
  },
  methods: {
    addOrder () {
      console.log(this.userId, this.quantity, this.status, this.complete)
      const userId = parseInt(this.userId)
      const quantity = parseInt(this.quantity)
      const status = this.status
      const complete = this.complete === 'true'
```

```
      this.$apollo.mutate({
        mutation: ADD_ORDER,
        variables: {
          userId,
          quantity,
          status,
          complete
        }
      })
      this.$router.push('/')
    }
  }
}
</script>
```

Also, include the new route at src/router/index.js:

```
import Vue from 'vue'
import Router from 'vue-router'
import HelloWorld from '@/components/HelloWorld'
import OrderById from '@/components/OrderById'
import AddOrder from '@/components/AddOrder'

Vue.use(Router)

export default new Router({
  routes: [
    {
      path: '/add',
      name: 'AddOrder',
      component: AddOrder
    },
    {
      path: '/',
      name: 'HelloWorld',
      component: HelloWorld
    },
    {
      path: '/:id',
      name: 'orderById',
      component: OrderById,
      props: true
    }
  ]
})
```

Go to the following URL:

```
http://localhost:8080/#/add
```

Then, create a new `order` with the parameters that are shown in the following screenshot and click on **Add**:

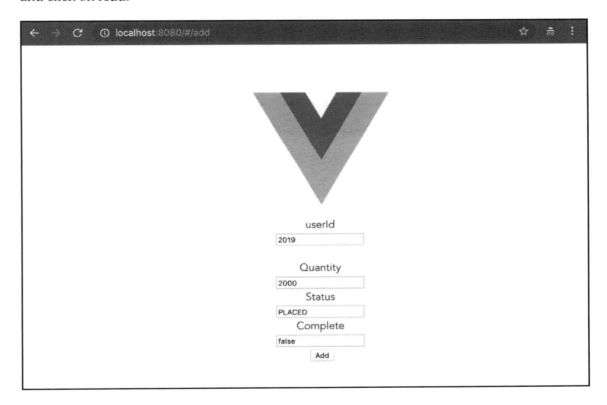

Creating a new order

You will then see the details of the newly created order:

Viewing the new order that was created

The following screenshot shows the created `order`:

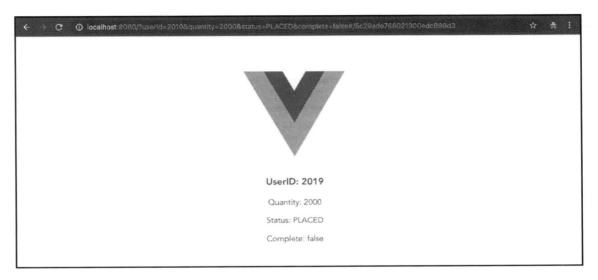

Viewing the single order that was created with GraphQL and Vue.js

Summary

This chapter presented GraphQL. You gained an understanding of the basics of this technology and learned how to change an existing API to support it.

You then learned how to manage GraphQL with Express.js and TypeScript, before moving on and learning about the playground interface, where you can test queries.

You were also able to create a single-page application with Vue.js to manipulate the `order` with GraphQL.

Questions

1. What is GraphQL?
2. Think of two reasons why you should use GraphQL.
3. Is GraphQL free?
4. Can GraphQL connect with an external data source?

5. What is GraphQL playground?
6. What do we mean by types in GraphQL?
7. What do we mean by mutations in GraphQL?
8. What is Vue.js?

Further reading

To improve your knowledge regarding what was covered in this chapter, the following books are recommended, as they will be helpful for the upcoming chapters:

- *Learning GraphQL and Relay* (https://www.packtpub.com/web-development/learning-graphql-and-relay)
- *Beginning GraphQL* (https://www.packtpub.com/application-development/beginning-graphql)
- *Vue.js Quick Start Guide* (https://www.packtpub.com/application-development/vuejs-quick-start-guide)

Assessments

Chapter 1

1. `PUT` can be used to update the entire resource and `PATCH` can be used to update specific information in a resource.
2. Using the `POST` method and passing the resource data as a body.
3. The REST architectural style constraints are as follows:

   - Uniform Interface
   - Stateless
   - Cacheable
   - Client-Server
   - Layered System
   - Code on Demand

4. Code on Demand.
5. `PATCH` method.
6. `PUT` method.
7. It is an HTTP method that's responsible for permitting the retrieval of a resource from the server.

Chapter 2

1. It means that this operation will retrieve a list of users from 200 until 219: 20 users.
2. The URL should look as follows:

```
https://<URI>/users?sort=first-name,last-name
```

3. The three main aspects of good resource naming are as follows:

   - **Understandability**: Both the server and the client should be able to understand and utilize the representation format of the resource.
   - **Completeness**: The format should be able to represent a resource completely.
   - **Linkability**: A resource can have a link to another resource.

4. No. The URI needs to always be written in the plural.
5. The recommendation is to keep the version in the URL, even though header-based versioning is more RESTful.
6. The five HTTP status code classes are as follows:

   - 1xx: Informational
   - 2xx: Success
   - 3xx: Redirection
   - 4xx: Client Error
   - 5xx: Server Error

7. **204**: No content.
8. **429**: Too Many Requests.

Chapter 3

1. API-First and Contract-First.
2. It means that it's time to start using HTTP verbs properly.
3. The OpenAPI is the official API specification. The specification is hosted by the OpenAPI Initiative and involves more than 30 organizations around the world including Microsoft, Google, IBM, and CapitalOne.
4. No, Swagger is a tool that has been built around the OpenAPI Specification. There are a lot of tools like Swagger that are also compliant with OpenAPI Spec.
5. The use of OpenAPI allows you to create a collaborative ecosystem on API design, in order to save time and avoid errors when writing code, ensure quality, and generate documentation.

6. Swagger is a set of open source tools built around the OpenAPI specification that can help you design, build, document, and consume REST APIs. The open source Swagger tools are Swagger Editor, Swagger UI, and Swagger Codegen.

7. Basically, be compliant with the RESTful concepts in practice, such as introducing resources, HTTP verbs, and hypermedia controls.

Chapter 4

1. Express.js, or simply Express, is a web application framework for Node.js.

2. npm is the package manager for the Node.js JavaScript platform. It puts modules in place so that the node can find them and manages dependency conflicts intelligently. It is extremely configurable and supports a wide variety of use cases.

3. Routing refers to how an application's endpoints (URIs) respond to client requests.

4. Visual Studio Code, or VSCode, is an open source and free source code editor developed by Microsoft for Windows, Linux, and macOS. It includes support for debugging, embedded Git control, syntax highlighting, intelligent code completion, snippets, and code refactoring.

5. A Linter or lint refers to tools that analyze source code to flag programming errors, bugs, stylistic errors, and suspicious constructs.

6. Under tsconfig.json file, add the following content:

```
{
  "compilerOptions": {
    "target": "es5",
    "module": "commonjs",
    "strict": true,
    "esModuleInterop": true,
    "outDir": "dist",
    "sourceMap": true
  }
}
```

7. From Mocha's website—Mocha is a feature-rich JavaScript test framework running on Node.js and in the browser, making asynchronous testing simple and fun.

8. Chai is a BDD/TDD assertion library for Node.js and the browser that can be delightfully paired with any JavaScript testing framework.

9. Stryker is a tool that provides ways to run mutation tests in Node.js applications.

Chapter 5

1. This file is responsible for maintaining the package's dependencies of all libraries the project is going to use.
2. You can point to them using `extends` in the `tslint.json` file, as follows:

```
{
  "extends": ["tslint:recommended", "tslint-config-prettier"],
  "rules": {
    "array-type": [true, "generic"],
    "no-string-literal": false,
    "object-literal-shorthand": [true, "never"],
    "only-arrow-functions": true,
    "interface-name": false,
    "max-classes-per-file": false,
    "no-var-requires": false,
    "ban-types": false
  }
}
```

3. Testing first generates a lot of benefits, such as the ability to anticipate acceptance criteria, better narrative understanding, and safer refactoring. This is because once a test is passing, it's safe to refactor it. Also, TDD results in more tests and consequently results in longer test run times and fewer bugs.
4. Postman is a Google Chrome app for interacting with HTTP APIs. It presents you with a friendly GUI for constructing requests and reading responses.
5. You should add the proper HTTP status code in the response methods.
6. It means that those dependencies will be used only in the `dev` environment.
7. `Mocha.opt` are properties you can define when Mocha is running, such as `--timeout 35000`, `--full-trace`, and so on.

Chapter 6

1. Before you write the code itself following TDD practices.
2. A model is the representation of a resource. For example, if we have a user resource, the user model could be as follows:

```
export interface User {
  id: Number
  username: String
  firstName: String
  lastName: String
```

```
        email: String
        password: String
        phone: String
        userStatus: Number
    }
```

3. Take the following URI:

```
http://localhost:3000/store/inventory?order-status=PLACED
```

You can get the query order-status as follows:

```
export let getInventory = (req: Request, res: Response, next:
NextFunction) => {
const status = req.query.status
...
}
```

4. You can pass unlimited query strings.
5. It should only contain route information and point the methods to the controllers that will handle and process the operation, as follows:

```
import * as orderController from '../controllers/order'

export class OrderRoute {
  public routes(app): void {
    app.route('/store/inventory').get(orderController.getInventory)
    app.route('/store/orders').post(orderController.addOrder)
    app.route('/store/orders').get(orderController.getAllOrders)
    app.route('/store/orders/:id').get(orderController.getOrder)
app.route('/store/orders/:id').delete(orderController.removeOrder)
    }
  }
}
```

6. This way, you can separate the responsibilities and make the tests easier.
7. It should ignore the offset and retrieve the maximum number of elements.

Chapter 7

1. Content negotiation refers to mechanisms defined as a part of HTTP that make it possible to serve different versions of a document at the same URI.
2. It is not required, but it is recommended.
3. HTTP Status **406**.

4. HAL, or Hypertext Application Language, is a standard convention for defining hypermedia such as links to external resources.
5. Yes. HAS is agnostic, and it doesn't matter if it is JSON or XML.
6. No, you can validate during the contract building and validation phase if necessary.
7. There are many libraries to do that—ketting, restl, hal-browser, hal-browser-express, and others.

Chapter 8

1. An **object data model (ODM)** is a data model based on object-oriented programming that associates methods (procedures) with objects that can benefit from class hierarchies.
2. In relational databases, you need to define constraints in terms of rows and named columns as well as the type of data that can be stored in each column. On the other hand, a document-oriented database contains documents, which are records that describe the data in the document.
3. Mongoose is an ODM library for MongoDB and Node.js.
4. A Mongoose *schema* is a document data structure (or shape of a document) that is enforced via the application layer.
5. The `ObjectId` class is the default primary key for a MongoDB document and is usually found in the `_id` field in an inserted document.
6. Docker is a computer program that performs operating-system-level virtualization, also known as *containerization*.
7. The command to spin up a Docker container with MongoDB is as follows:

```
docker run --name my-mongo -p 27017:27017 -v /data/mongo:/data/db -d mongo:latest
```

8. The `save` method is as follows:

```
const newUser = new UserModel(req.body)

newUser.save((error, user) => {...})
```

9. The `findById` method is as follows:

```
UserModel.findById(userId, (err, user) => {
...
}
```

Chapter 9

1. It means a process that checks whether the requestor has the credentials needed to get access to the application.
2. It means a process that checks whether the requestor has the rights needed to execute a specific piece of the application.
3. JSON Web Token is a JSON-based open standard for creating access tokens that assert some claims.
4. bcrypt is a password hashing function designed by Niels Provos and David Mazières, based on the Blowfish cipher, and presented at USENIX in 1999.
5. `mongoose-unique-validator` is a plugin that adds pre-save validation for unique fields within a Mongoose schema.
6. HTTPS offers an extra layer of security because it uses SSL to move data.
7. It is not. You should use any other strategy, such as Cloud Storage, encryption, and/or vault. To reduce the complexity, we can decide to put the secret in the code base only because it is not production code.

Chapter 10

1. An error handler is used to catch errors in an application, and logging is a strategy used to grab pieces of information throughout runtimes, such as errors or relevant information.
2. You can define them all at the end of the `app.use` definition, as follows:

```
import * as bodyParser from 'body-parser'
import * as express from 'express'
import * as mongoose from 'mongoose'
import { APIRoute } from '../src/routes/api'
import { OrderRoute } from '../src/routes/order'
import { UserRoute } from '../src/routes/user'
import * as errorHandler from '../src/utility/errorHandler'

class App {
  public app: express.Application
  public userRoutes: UserRoute = new UserRoute()
  public apiRoutes: APIRoute = new APIRoute()
  public orderRoutes: OrderRoute = new OrderRoute()
  public mongoUrl: string = 'mongodb://localhost/order-api'

  constructor() {
    this.app = express()
```

```
      this.app.use(bodyParser.json())
      this.userRoutes.routes(this.app)
      this.apiRoutes.routes(this.app)
      this.orderRoutes.routes(this.app)
      this.mongoSetup()
      this.app.use(errorHandler.logging)
      this.app.use(errorHandler.clientErrorHandler)
      this.app.use(errorHandler.errorHandler)
    }

    private mongoSetup(): void {
      mongoose.connect(
        this.mongoUrl,
        { useNewUrlParser: true }
      )
    }
  }

  export default new App().app
```

3. Yes, the error handler definitions must be added at the end, otherwise they won't catch the errors.

4. It is recommended that you log every operation of an application. This log strategy is more involved than logging only when errors occur. It is also used for logging critical operations, for instance, so that you can have more control over who is doing what for audit purposes. Applications can log at the code level (debugging, for instance) and at the user request level for audits and checking access.

5. Yes, there are a lot of ways, and one of them is adding the application level definition as follows:

```
import * as bodyParser from 'body-parser'
import * as express from 'express'
import * as expressWinston from 'express-winston'
import * as mongoose from 'mongoose'
import * as winston from 'winston'
import { APIRoute } from '../src/routes/api'
import { OrderRoute } from '../src/routes/order'
import { UserRoute } from '../src/routes/user'
import * as errorHandler from '../src/utility/errorHandler'

class App {
  public app: express.Application
  public userRoutes: UserRoute = new UserRoute()
  public apiRoutes: APIRoute = new APIRoute()
  public orderRoutes: OrderRoute = new OrderRoute()
```

```
      public mongoUrl: string = 'mongodb://localhost/order-api'

      constructor() {
        this.app = express()
        this.app.use(bodyParser.json())
        this.userRoutes.routes(this.app)
        this.apiRoutes.routes(this.app)
        this.orderRoutes.routes(this.app)
        this.mongoSetup()
        this.app.use(
          expressWinston.errorLogger({
            transports: [new winston.transports.Console()],
          })
        )
        this.app.use(errorHandler.logging)
        this.app.use(errorHandler.clientErrorHandler)
        this.app.use(errorHandler.errorHandler)
      }

      private mongoSetup(): void {
        mongoose.connect(
          this.mongoUrl,
          { useNewUrlParser: true }
        )
      }
    }

    export default new App().app
```

6. You can use the next() function, as follows:

```
    import { NextFunction, Request, Response } from 'express'
    import { OrderAPILogger } from '../utility/logger'

    export let logging = (
      err: Error,
      req: Request,
      res: Response,
      next: NextFunction
    ) => {
      OrderAPILogger.logger.error(err)
      next(err)
    }

    export let clientErrorHandler = (
      err: Error,
      req: Request,
      res: Response,
```

```
    next: NextFunction
) => {
  if (req.xhr) {
    res.status(500).send({ error: 'Something failed!' })
  } else {
    next(err)
  }
}

export let errorHandler = (
  err: Error,
  req: Request,
  res: Response,
  next: NextFunction
) => {
  res.status(500).send({ error: err.message })
}
```

7. It passes the error forward to the application.

Chapter 11

1. The term **Continuous Integration** (**CI**) refers to a development practice in which developers can integrate their code into a shared repository several times per day. Every time code is sent to a repository, an automated process runs several processes, such as an automated build, testing, and other tasks that help developers to identify possible problems before going to production.

2. **Continuous Delivery** (**CD**) is a software release approach in which development teams produce and test code in short cycles and go more quickly to production every time when there is at least one green build.

3. Travis CI is a hosted, distributed CI service used to build and test software projects hosted in GitHub.

4. **Google Cloud Platform** (**GCP**) is a suite of cloud computing services that run on the same infrastructure that Google uses internally for its end-user products, such as Google Search and YouTube.

5. There are a few possibilities, as described in Google's documentation:

   - Create a Google Compute Engine virtual machine with MongoDB pre-installed.
   - Create a MongoDB instance with MongoDB Atlas on GCP.
   - Use mLab to create a free MongoDB deployment on Google Cloud Platform.

6. Sensitive information, such as credentials.

7. This new version won't be available for production.

Chapter 12

1. No, it is more complex than monolithic applications due to the ecosystem involved in the process and the domain responsibilities.

2. Simple to test, scale, and deploy.

3. Each microservice should have its own database enfolding its domain.

4. Yes, Continuous Integration is a part of microservice architecture.

5. Like CI, CD is used in microservice architectures to promote changes to production more quickly.

6. Yes. Your ecosystem should allow you to use any language you want. The microservice has to expose the business capability without being associated with any programming language.

7. A developer portal is an interface between a set of APIs and their various stakeholders.

Chapter 13

1. GraphQL is an open source data query and manipulation language for APIs, and a runtime for fulfilling queries with existing data.

2. GraphQL enables the possibility of avoiding over-fetching data and enables rapid product development.

3. Yes. It is open source under Facebook's license, and is available at `https://opensource.fb.com`.

4. Yes. You can get the data from wherever you want and use GraphQL.

5. GraphQL Playground is a graphical, interactive, in-browser GraphQL IDE, created by Prisma and based on GraphQL.

6. The most basic components of a GraphQL schema are object types, which represent a kind of object you can fetch from your service, and what fields it has. The following is an example:

```
type Character {
  name: String!
  appearsIn: [Episode!]!
}
```

7. It is when something might change the data state, such as creating a new object or updating an existing one, as follows:

```
export default new GraphQLSchema({
  query: QueryType,
  mutation: MutationType
});
```

8. Vue.js is an open source JavaScript framework for building user interfaces and single-page applications.

Other Books You May Enjoy

If you enjoyed this book, you may be interested in these other books by Packt:

Mastering TypeScript 3 - Third Edition
Nathan Rozentals

ISBN: 9781789536706

- Gain insights into core and advanced TypeScript language features
- Integrate existing JavaScript libraries and third-party frameworks using declaration files
- Target popular JavaScript frameworks, such as Angular, React, and more
- Create test suites for your application with Jasmine and Selenium
- Organize your application code using modules, AMD loaders, and SystemJS
- Explore advanced object-oriented design principles
- Compare the various MVC implementations in Aurelia, Angular, React, and more

Hands-On TypeScript for C# and .NET Core Developers
Francesco Abbruzzese

ISBN: 9781789130287

- Organize, test, and package large TypeScript code base
- Add TypeScript to projects using TypeScript declaration files
- Perform DOM manipulation with TypeScript
- Develop Angular projects with the Visual Studio Angular project template
- Define and use inheritance, abstract classes, and methods
- Leverage TypeScript-type compatibility rules
- Use WebPack to bundle JavaScript and other resources such as CSS to improve performance
- Build custom directives and attributes, and learn about animations

Leave a review - let other readers know what you think

Please share your thoughts on this book with others by leaving a review on the site that you bought it from. If you purchased the book from Amazon, please leave us an honest review on this book's Amazon page. This is vital so that other potential readers can see and use your unbiased opinion to make purchasing decisions, we can understand what our customers think about our products, and our authors can see your feedback on the title that they have worked with Packt to create. It will only take a few minutes of your time, but is valuable to other potential customers, our authors, and Packt. Thank you!

Index

testing 94, 96, 98

O

Object Data Modeling (ODM) 251
One-Size-Fits-All (OSFA) 27
OpenAPI Specification (OAS)
　about 47
　data types 51
　document structure 51
　format 50
OpenSSL
　reference 280
Order API application
　file structure 142, 143, 145
　initial configurations 142, 143, 145
　routes, controlling 146, 147, 148
　routes, defining 146, 147, 148, 150
　serving 142
Order Management System (OMS) 141
order-api, setting up with Mongo
　controllers, modifying 262, 265, 269
　schemas, creating 253
　tests, changing 255, 258, 262
order-api
　breaking down 371
　databases, creating 373
　error handler, adding 308, 310
　GCP projects 377
　projects, creating on GitHub 375
　setting up, with Mongo 251
　skeleton, creating for microservices 386
　swagger implementation 377, 381, 385

P

pagination
　implementing 180, 182, 191, 195, 196
Postman
　reference 150
　used, for testing 150
Prettier 120, 121

Q

query strings
　implementing 180, 182

R

Representational State Transfer (REST)
　about 9, 10
　architectural styles 10
resource URIs
　controllers, implementing 163, 167, 168
　creating 154, 156
　executing, with Postman 171, 173, 175, 177, 178
　order model, creating 156
　remaining routes, configuring 169, 170
　tests, creating for missing routes 158, 160, 163
　user model, creating 156
resource-based APIs 26
RESTful services
　HTTP methods 19, 22
Richardson Maturity Model (RMM)
　design 52
　HTTP Verbs (level 2) 56
　Hypermedia Controls (level 3) 58, 60
　implementation 52
　Resources (level 1) 54
　Swamp of POX (level 0) 52
Robomongo
　about 248, 250
　reference 248

S

security
　HTTPS, using 280
　overview 280
　testing, manually through Postman 299, 301, 302
Simple Object Access Protocol (SOAP) 10
skeleton, microservic
　user-ms code, using 404
skeleton, microservice
　creating 386
　order-ms code structure 386, 388, 390, 392, 395, 397
　user-ms code, using 400
Stryker
　about 126
　used, for testing mutation tests 126, 128, 131,

Made in the
USA
Middletown, DE